WOODY BOG PLANTS

A Field Guide for New England and Adjacent Canada

Mountain Holly
(*Ilex mucronata*)

STEVE CHADDE, FAY HYLAND,
and BARBARA HOISINGTON

WOODY BOG PLANTS
A Field Guide for New England and Adjacent Canada

Steve Chadde, Fay Hyland, and Barbara Hoisington

A Pathfinder Field Guide
Published by Orchard Innovations, Mountain View, Arkansas

FRONTIS ILLUSTRATION: Mountain Holly (*Ilex mucronata*), drawing by Laurel Smith.
The author may be contacted at: *steve@orchardinnovations.com*

CONTENTS

    (*including both scientific and common names*)

JUSTIN HENRY

Typical New England bog. Stunted trees of black spruce (*Picea mariana*) and members of the Heath family (Ericaceae) dominate the woody flora.

INTRODUCTION

Bogs are fascinating places to visit! These wetlands, dominated by various sphagnum mosses (see p. 112), are covered with a wall-to-wall, undulating carpet of this moss. To the lay-person, these areas might appear as monotonous assemblages of only a few species of stunted plants so similar in appearance as to appear homogeneous, but a closer look will detect as many different species as might be found in a rich woods. By use of the keys, descriptions, and photographs in this guide, more than fifty different kinds of woody bog plants of the New England region (and adjacent Canada) can be readily identified.

Sphagnum moss, orchids, cottongrass, and carnivorous plants (pitcher plants, sundews, bladderworts) are conspicuous non-woody components of the bog mat, but by far the bulk of the tough fibrous material which binds the structure together is composed of the living and partially decayed roots and small stems of woody plants. Decay is slow due to the high acidity (pH about 4) and low oxygen content of the substrate. It is questionable whether living roots are present in any but the uppermost layers of the bog, the plants depending upon the relatively less acid rain water being held by the upper layers of sphagnum.

Compared to well-drained areas, the organic matter of bogs decomposes (mineralizes) extremely slowly due to the unfavorable conditions for microbial action. Available nitrogen and other nutrients are insufficient for vigorous growth of much of the bog flora, especially so of the larger woody plants which often develop into dwarfed and unhealthy individuals. Dead and dying trees of cedar, tamarack, and spruce are common. Much of the black spruce population results from rooting of the lower branches. Although cones with viable seeds are produced, the conditions necessary for germination and survival are often lacking in the sterile sphagnum mat which surrounds these trees.

Most of the woody species included here may also be found in habitats where growing conditions are more favorable, but these particular plants have been singularly successful in adapting to the adverse conditions found in bogs. Several species which may not strictly be considered typical bog plants, but which usually are encountered along the wet margins of bogs, often invading the sphagnum mat, have been included.

The bogs are continuously wet and support a luxurious surface growth of living sphagnum moss (*Sphagnum* spp.) which carpets the area and which, upon decay, adds considerably to the peaty material below the surface.

Based on radio carbon dating and other methods, these New England bogs are variously considered to be between 7,000 and 11,000 years old. They may exceed 30 feet in depth in some places. Studies of pollen grains of plants which grew in the area during development of the bogs reflects not only the types of plants present from time to time, but also the climates of the recent past.

The sphagnum-dominated areas of bogs often merge imperceptibly into swamps or open water on the side bordered by low, flat terrain. In these areas, heath plants and several others are replaced by cattails, skunk cabbage, willows, grasses, sedges, rushes, and other typical aquatics, but the trees may persist. On the side of the bog bordered by a steep slope, the flora merges abruptly into the mesophytic type characteristic of the local area. Sphagnum is eliminated as light and moisture are reduced, and mineral soil becomes available to the plants.

This bulletin is an illustrated guide to the identification of the woody flora commonly composing the fibrous mat of the typical continuously wet, acid, sphagnum-covered low moor type of bog, which is so common in the northeastern United States and adjacent Canada. High moor and alkaline bogs support a somewhat similar flora, but there are enough differences from the above to warrant a separate study.

Scientific and Common Names

Plants are arranged alphabetically in this guide, first by the scientific name of the family, then within the family by their scientific (or latin) name. The scientific names follow those of the **Plants of the World Online**, a database of the world's flora maintained by the Royal Botanic Gardens—Kew (see *powo.science.kew.org*). The common or vernacular names used in this bulletin were obtained from a variety of sources. Only one common name is used in conjunction with the scientific name accompanying the photographs and text, but several others, if relevant, are included at the end of each species' description, and all are listed in the index. Since many Canadian plants are featured in the guide, the common names used in Quebec and Newfoundland are also provided.

About this Guide

Much of the text of this guide was originally published in 1977 by the Life Sciences and Agriculture Experiment Station, University of Maine, as Bulletin 744, and entitled: *The Woody Plants of Sphagnous Bogs of Northern New England and Adjacent Canada*. Authors were **Fay Hyland**, Professor Emeritus, Department of Botany and Plant Pathology, University of Maine at Orono, and **Barbara Hoisington**, at the time, a botany student, University of Maine at Orono.

This new edition has been revised and updated to reflect current family assignments and species nomenclature, plus adds county-level distribution maps for the New England region, a North American distribution map, and photographs for each species described.

Acknowledgment is given to the **Biota of North America Program** (BONAP) for use of their data to generate the regional maps, and the **Royal Botanic Gardens—Kew** for use of their world-wide distribution maps. Grateful acknowledgment is also given to the many photographers who have made their images available under a Creative Commons commercial-use license. Photographer names are indicated on each photograph. Photographs were obtained primarily from the *inaturalist.org* website, an invaluable resource documenting occurences of the world's flora and fauna.

Conservation

Bogs are a unique feature of the environment, and are home to many rare plant species. Care should be used when visiting these special wetlands as the sphagnum hummocks and mats are easily damaged by trampling, and are slow to recover. The outer edge of the bog is often a moat-like ring of deep water, and may be difficult to cross onto the body of the bog. When on the bog and attempting to identify an unknown plant, do not pick plant parts, but rather observe the plant at its level. With practice, characteristics of plant families will become familiar, making identification quicker and easier.

For identification of non-woody bog plants, and plants, in general, of the New England region, a list of selected references is provided on page 8.

How to Use the Key

With an unknown specimen before you, read both leads numbered 1 in the key and decide which best fits your plant. If it is the first number 1, then go on to the pair of leads numbered 2, and again decide which is the correct choice. However, if the second number 1 fits your specimen better, then proceed with the number indicated at the end of the lead (number 7 in the case of the following key). Keep on proceeding through the key, always choosing the number of each pair which fits your specimen, until you come to the name of the plant (common name in red, followed by the scientific name in *italics*). Now turn to the page indicated and check your specimen with the description and photographs.

Another (but less satisfactory) method is simply to leaf through the guide until you find a specimen which matches yours, then read the description on the facing page to see if it fits. A metric ruler is printed on the back page which may be used to measure parts of specimens.

SELECTED REFERENCES

Books

Aquatic and Wetland Plants of Northeastern North America. Volumes 1 & 2. 2006. Garrett E. Crow and C. Barre Hellquist. University of Wisconsin Press.

Bogs & Fens: A Guide to the Peatland Plants of the Northeastern United States and Adjacent Canada. 2016. Ronald B. Davis. University Press of New England.

Field Guide to Eastern Trees: Peterson Field Guide. 1998. George A. Petrides. Houghton Mifflin Company.

Field Guide to Wildflowers: Northeastern and North-Central North America: Peterson Field Guide. 1998. Roger Tory Peterson and Margaret McKenny. Houghton Mifflin Company.

Flora of the Northeast: A Manual of the Vascular Flora of New England and Adjacent New York. 2nd ed. 2007. Dennis W. Magee and Henry E. Ahles. University of Massachusetts Press.

Meet the Peat: A field guide produced for the Josselyn Botanical Society. 2024. Ralph Pope.

New England Wild Flower Society's Flora Novae Angliae: A Manual for the Identification of Native and Naturalized Higher Vascular Plants of New England. 2011. Arthur Haines.

Newcomb's Wildflower Guide. 1977. Lawrence Newcomb. Little, Brown and Company.

Northeast Ferns A Field Guide to the Ferns and Fern Relatives of the Northeastern United States. 2022. Steve Chadde. Orchard Innovations.

Wetland Plants of New England: A Guide to Trees, Shrubs, and Lianas. 2016. Dr. Donald J. Padgett. Spatterdock Press.

Wildflowers in Field and Forest: A Field guide to the Northeastern United States. 2006. Steven Clemants and Carol Gracie. Oxford University Press.

Wildflowers of Maine, New Hampshire, and Vermont. 2001. Arleen R. Bessette, William K. Chapman, Valerie A. Chapman and Alan E. Bessette ed. Syracuse University Press.

Websites

iNaturalist: *inaturalist.org*
Native Plant Trust: *gobotany.nativeplanttrust.org*
Plants of the World Online: *powo.science.kew.org*

KEY TO THE SPECIES

1 Seeds borne naked in woody or berry-like cones. Leaves evergreen (except in **Tamarack**), awl-shaped, scale-like or linear. Trees, mostly resinous. . . . . **Gymnosperms, 2**

1 Seeds borne in an ovary. Leaves deciduous (except those of several low shrubs), mostly broad and flat . . . . **Angiosperms, 7**

  2 Leaves needle-like, borne in fascicles of 2, bound together at the base by a membranaceous sheath; cones over 2.5 cm long . . . . **Jack Pine** (*Pinus banksiana*) (p. 82)

  2 Leaves linear, awl-shaped or scale-like, without a membranaceous sheath at base; cones less than 2.5 cm long 3

3 Leaves linear, alternate 4

3 Leaves scale-shaped or awl-shaped, opposite, or in threes 5

  4 Leaves deciduous, soft, flexible, borne in clusters of 8 to many on the spur branches, but scattered singly along the leading shoots. Cones erect . . . . **Tamarack** (*Larix laricina*) (p. 78)

  4 Leaves evergreen, stiff, pointed, spirally arranged on all the branches. Cones pendent. . . . . **Black Spruce** (*Picea mariana*) (p. 80)

5 Branchlets rounded. Cone bluish or whitened, fleshy and berrylike . . . . **Eastern Red-cedar** (*Juniperus virginiana*) (p. 32)

5 Branchlets flattened. Cone composed of leathery or slightly woody scales 6

  6 Branchlets much flattened. Cones longer than broad, composed of flat, uniformly thickened, tan or brown scales . . . . **Northern White-cedar** (*Thuja occidentalis*) (p. 34)

  6 Branchlets slightly flattened. Cones rounded, composed of bluish or whitened wedge-shaped or tack-shaped scales thickened at their outer ends and narrow toward the central point where they are attached . . . **Atlantic White-cedar** (*Chamaecyparis thyoides*) (p. 30)

7 Tiny plant, parasitic on the branches of Picea and Larix . . . . **Dwarf Mistletoe** (*Arceuthobium pusillum*) (p. 104)

7 Larger plants, not parasitic on the branches of conifers 8

  8 Leaves persistent, evergreen 9

  8 Leaves deciduous 18

9 Leaves opposite or whorled. 10

9 Leaves alternate. 11

  10 Leaves either opposite or whorled, older leaves bending down on curved petioles. Twigs circular in cross section. Fruit axillary on slender, curved pedicels . **Sheep-Laurel** (*Kalmia angustifolia*) (p. 48)

  10 Leaves all opposite, not bending down on curved petioles. Twigs 2-edged. Fruit terminal on long, slender, erect pedicels . . . . **Bog-Laurel** (*Kalmia polifolia*) (p. 50)

10 ■ KEY TO THE SPECIES

21 Leaves pubescent with persistent thin, dry, sheathing stipules. Leaflets 5–7, margins revolute. Fruit an achene. Low much branched shrub . . . .**Shrubby Cinquefoil** (*Dasiphora fruticosa*) (p. 90)

21 Leaves glabrous, without persistent thin, dry, sheathing stipules. Leaflets 7–13, margins not revolute. Fruit a drupe. Small tree . . . .**Poison Sumac** . . . . (*Toxicodendron vernix*) (p. 14)

22 Leaves lobed . . . .**Red Maple** (*Acer rubrum*) (p. 106)

22 Leaves not lobed **23**

23 Leaves opposite or whorled. **24**

23 Leaves alternate **29**

24 Leaves mostly whorled **25**

24 Leaves all opposite **26**

25 Stems recurved, arching and rooting at the tips, the submersed ones spongy-thickened. Leaves lanceolate. Fruit not in heads **Water-willow** . . . . (*Decodon verticillatus*) (p. 70)

25 Stems upright, neither arching nor rooting at the tips, not spongy-thickened. Leaves ovate. Fruit in heads . . . .**Buttonbush** . . . . (*Cephalanthus occidentalis*) (p. 100)

26 Leaves entire and oblong. Fruit a berry. Branchlets brittle and weak **27**

26 Leaves at least remotely serrate or dentate, elliptic-oblong, elliptic, or ovate. Fruit a drupe. Branchlets stiff and strong **28**

27 Leaves tapering at base, glabrous or nearly so. Bark grayish. Berry red or purple . . . .**Swamp Fly-Honeysuckle** (*Lonicera oblongifolia*) (p. 26)

27 Leaves rounded at base, hairy on both surfaces, ciliate. Bark or branchlets brown. Berry bluish-black . . . .**Mountain Fly-Honeysuckle** . . . . (*Lonicera villosa*) (p. 28)

28 Leaves elliptic-oblong, margin wavy or remotely toothed. Twigs terete. . . . .**Witherod** (*Viburnum cassinoides*) (p. 108)

28 Leaves ovate and dentate. Twigs angled. . . . . **Smooth Arrow-wood** . . . . (*Viburnum recognitum*) (p. 110)

29 Leaves entire **30**

29 Leaves serrate or crenate **38**

30 Pith of twigs diaphragmed. Large tree. . . . .**Black Tupelo** . . . . (*Nyssa sylvatica*) (p. 74)

30 Pith of twigs homogeneous. Shrub. **31**

31 Bud with a single cap-like scale . . **Bog Willow** (*Salix pedicellaris*) (p. 102)

31 Buds with more than one bud scale. **32**

32 Leaves mucronate; petioles and young twigs purple . . . .**Mountain Holly** (*Ilex mucronata*) (p. 18)

32 Leaves not mucronate; petioles and young twigs not purple **33**

33 Fruit a berry or berry-like drupe **34**

33 Fruit a capsule **37**

34 Leaves sprinkled with resinous dots (sticky when pinched between thumb and finger). Fruit a berry-like drupe containing ten seed-like nutlets **35**

34 Leaves not sprinkled with resinous dots (not sticky). Fruit a berry with more than ten tiny seeds **36**

35 Inflorescence with conspicuous leaf-like bracts . . . **Dwarf Huckleberry** . . . . (*Gaylussacia bigeloviana*) (p. 46)

35 Inflorescence without conspicuous leaf-like bracts. **Black Huckleberry** . . . . (*Gaylussacia baccata*) (p. 44)

36 Shrub 1–4 m high. Leaves downy or woolly underneath. Berries polished black, without bloom . . . . **Black Highbush** Blueberry . . . . (*Vaccinium fuscatum*) (p. 62)

36 Shrub 2–6 dm high. Leaves downy on both sides. Berries, blue with much bloom (rarely black) . . . . **Velvet-leaf Blueberry** . . . . (*Vaccinium myrtilloides*) (p. 66)

37 Capsule much longer than broad . . . . **Rhodora** . . . . (*Rhododendron canadense*) (p. 54)

37 Capsule globular . . . . **Maleberry** (*Lyonia ligustrina*)

38 Bud with a single cap-like scale . . . . **Willow** (*Salix* spp.).*
*Although not strictly bog plants, several species of willow are common in areas bordering the bog where there is enough mineral matter to support their growth. These species occasionally encroach upon the bog and become established where the sphagnous mat is sparse or lacking. A taxonomy manual should be consulted for species identification.

38 Buds with more than one bud scale **39**

39 Leaves doubly serrate. Pith triangular in cross section . **Speckled Alder** . . . . (*Alnus incana*) (p. 22)

39 Leaves not doubly serrate. Pith not triangular in cross section **40**

40 Leaves resin-dotted, aromatic, entire below the middle but with few coarse teeth near apex. . . . . **Sweet Gale** (*Myrica gale*) (p. 72)

40 Leaves neither resin-dotted nor aromatic, serrate to the base . . . **41**

41 Leaves with glands on upper surface of midrib of leaf near petiole . . **42**

41 Leaves without glands on upper surface of midrib of leaf near petiole **43**

42 New branchlets, leaves, rachis, and pedicels more or less densely gray or white-tomentose. Leaves varying from broadly oblanceolate to narrowly obovate or subelliptic, abruptly short acuminate at apex, tapering to the base. Fruit purple or purple-black . . . . **Purple Chokeberry** (*Aronia prunifolia*) (p. 88)

42 New branchlets, leaves, rachis and pedicels essentially glabrous. Leaves variable but tending to be less obovate and abruptly pointed at tips Fruit black . . **Black Chokeberry** (*Aronia melanocarpa*) (p. 86)

43 Leaf blades nearly as broad as long. Fruit a conelike aggregate of tiny nutlets or samaras . . . . **Bog Birch** (*Betula pumila*) (p. 24)

43 Leaves narrow, much longer than broad. Fruit not a conelike aggregate of tiny nutlets **44**

44 Fruit a capsule . . . . **Maleberry** (*Lyonia ligustrina*) (p. 52)

44 Fruit not a capsule **45**

45 Fruit a berry or berry-like drupe **46**

45 Fruit a follicle **50**

46 Fruit a berry-like drupe with few, rather large seeds **47**

46 Fruit a berry with many tiny seeds **49**

47 Fruit black. Shrub with few branches, or not branched . . . .
 . . . . **Alder-leaf Buckthorn** (*Rhamnus alnifolia*) (p. 84)

47 Fruit red. Shrub much branched **48**

48 Leaves thick, brittle, nearly glabrous above, pubescent, at least on the veins beneath. Fruit nearly sessile. Common and widespread . .
 . . . . **Common Winterberry** (*Ilex verticillata*) (p. 20)

48 Leaves thin, glabrous on both sides. Fruit not sessile; length of pedicels about equaling the diameter of the fruit . . . .
 . . . . **Smooth Winterberry** (*Ilex laevigata*) (p. 16)

49 Shrub less than 0.5 dm high, much branched, freely stoloniferous
 . . . . **Low Sweet Blueberry** (*Vaccinium angustifolium*) (p. 58)

49 Shrub to 4 m high forming compact or open clumps . . . .
 . . . . **Highbush Blueberry** (*Vaccinium corymbosum*) (p. 60)

50 Leaves glabrous, broadly oblanceolate or obovate. Twigs glabrous, tan-brown . . . . **Meadowsweet** (*Spiraea latifolia*) (p. 96)

50 Leaves tomentose, broad elliptic to elliptic-oblong. Twigs tomentose, purplish . . . . **Hardhack** (*Spiraea tomentosa*) (p. 98)

SHIRLEY ZUNDELL

■ Poison-Sumac

Toxicodendron vernix (L.) Kuntz

GROWTH FORM A coarse shrub or small tree 2–7 m high, with gray, smooth bark and glabrous, glaucous branchlets.

LEAVES Deciduous, alternate, odd-pinnately compound with 7–13 entire, oblong-obovate to elliptic, acuminate, shiny leaflets. The general appearance of the plant, with its compound leaves and reddish petioles, closely resembles the American Mountain-ash (*Sorbus americana*) but may be distinguished by its entire leaflets. (Mountain-ash leaflets are serrate).

FLOWERS Greenish-yellow, small, borne in slender axillary panicles 8–20 cm long, May–June.

FRUIT A subglobose or compressed, whitish or drab, dry drupe, August to November.

WHERE FOUND Bogs, especially near the borders where some mineral soil is available; also in wooded swamps.

NOTES The rather attractive whitish fruits may persist on the twigs all winter and are sometimes tempting to the novice for floral displays, but if any part of the plant is touched, it may cause a mild to severe skin rash. Bobwhites, pheasants, and grouse consume the fruits in winter.

OTHER NAMES Poison-dogwood, Poison-elder.

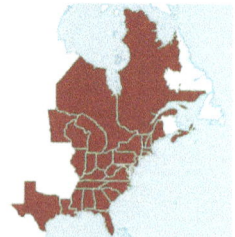

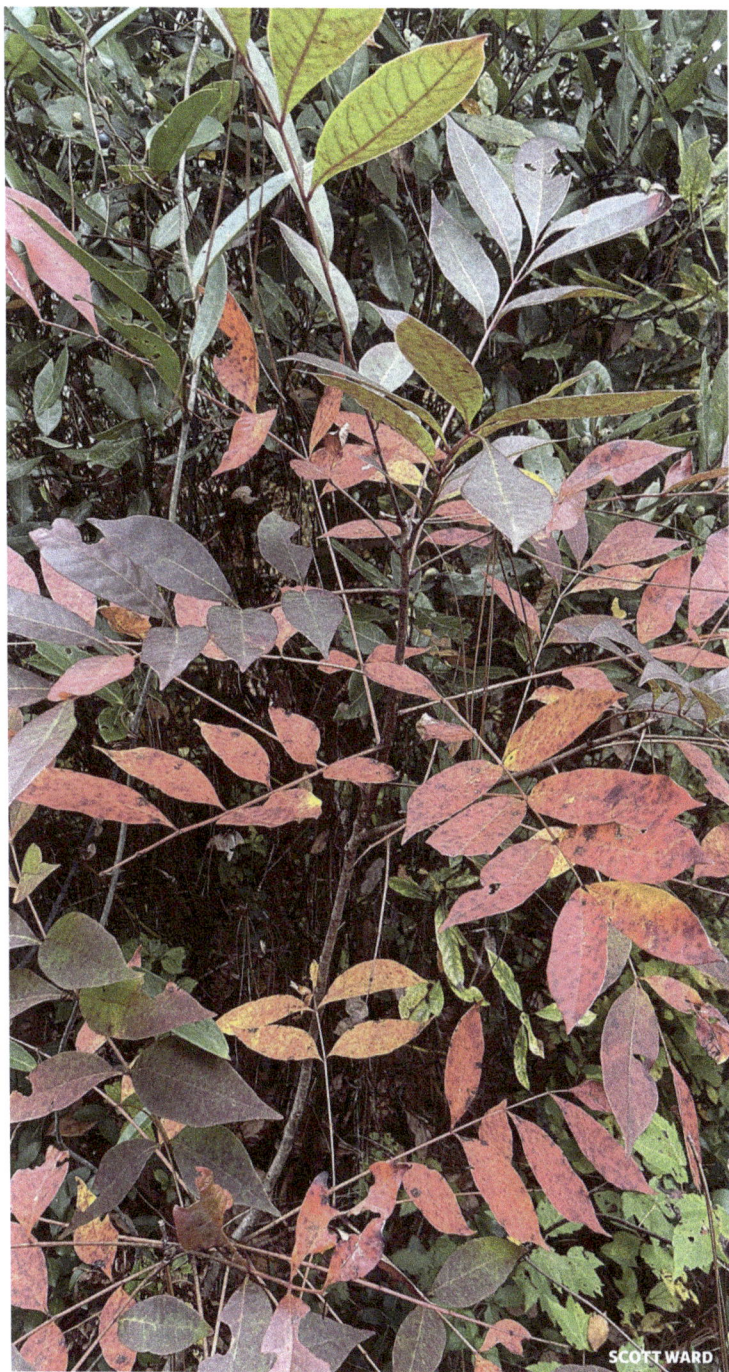

SCOTT WARD

POISON-SUMAC (*Toxicodendron vernix*)

KEVIN KENNY

■ Smooth Winterberry

Ilex laevigata (Pursh) Gray

GROWTH FORM Shrub to 4 m tall, with upright, glabrous branches.

LEAVES Deciduous, alternate, simple, lance-ovate to lanceolate or oval-elliptic, 3–9 cm long, lustrous above, glabrous or pilose on the veins beneath, appressed-serrulate, long acuminate, turning yellow in the fall.

FLOWERS Small, with 4 or 5 white petals, borne on short stalks, May–June.

FRUIT A depressed-globose, scarlet or orange-red (rarely bright yellow) drupe, 7–8 mm in diameter, with persistent calyx, solitary, on stalks 2–5 mm long, persistent.

WHERE FOUND Thin, sphagnous areas in the bog but more often in wooded swamps or peaty areas.

NOTES Sometimes planted for its orange-red fruits (which are also a food for birds). Absent north of southern Maine.

OTHER NAME *Aulne blanche* (Quebec).

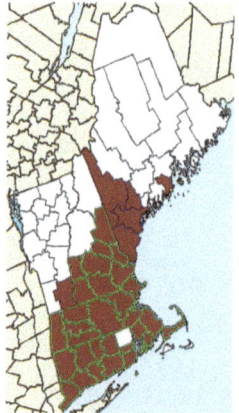

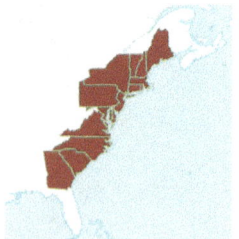

BONNIE SEMMLING

SMOOTH WINTERBERRY (*Ilex laevigata*)

IAN MANNING

■ Mountain Holly

Ilex mucronata (L.) Powell, Savolainen & S. Andrews

GROWTH FORM Erect, slender-branched shrub, 0.3–3 m high with purplish young branchlets and ashy-gray older bark.

LEAVES Deciduous, simple, alternate, entire (or sometimes with a few coarse teeth), smooth, thin, elliptic-oblong to narrowly obovate, mucronate, 2.5–3.5 cm long, turning yellow in autumn; borne on purple petioles, 6–12 mm long.

FLOWERS Small, whitish, 4–5 mm across, borne on slender, axillary, pedicels 1–3 cm long, May–June.

FRUIT A dull, subglobose, red drupe 6–8 mm across, containing 4–5 bony nutlets; pedicels 1–3 cm long.

WHERE FOUND Open areas in the bog where the sphagnous mat is thin, but making better development in damp woods, swamps, and uplands.

NOTES A common shrub often unfamiliar to the layperson. Mountain Holly is not as conspicuous in fruit as is the **Common Winterberry**, *Ilex verticillata,* with which it often grows. Mountain Holly is a host plant for the caterpillars of the **Columbia silkmoth** (*Hyalophora columbia*) found in Maine, which has an impressive 3.5-inch (10 cm) wingspan.

OTHER NAMES Catberry, *Faux Houx* (Quebec). Formerly classified as *Nemopanthus mucronata* (L.) Trel.

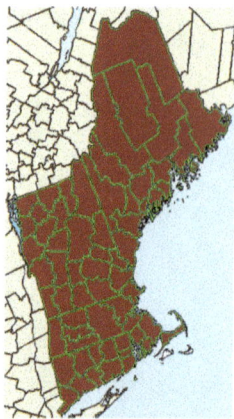

MOUNTAIN HOLLY (*Ilex mucronata*)

DAWN DAVIS

■ Common Winterberry

Ilex verticillata (L.) Gray

GROWTH FORM Common shrub 1–4 m high with spreading branches (occasionally upright).

LEAVES Deciduous, alternate, simple, lanceolate to oblong-lanceolate or obovate, rather stiff, firm, and thick, sharply serrate or doubly serrate with callous (somewhat spinescent) teeth; appressed-pilose or glabrous beneath, dull above, 3–7 cm long, turning black or dull yellowish-brown after frost.

FLOWERS Small, with 5–8 white petals, borne on short stalks, June–August.

FRUIT A bright red (rarely yellow) globose drupe, 5–7 mm in diameter, with persistent calyx, borne on short stalks, arranged in false whorls (verticillate) and persistent on the branchlets for several weeks.

WHERE FOUND Open places in the bog and along the margins, often growing in water; also found in swamps, damp thickets, and along pond-margins.

NOTES Highly decorative, valued for its bright red fruits which persist for a long time on the branchlets and are a valuable winter food for birds. When abundant, the conspicuous fruits add a pleasing color to the landscape. Several varieties are recognized by botanists.

OTHER NAMES *Apalanche* (Quebec), *Aulne blanche* (Quebec), **Black Alder, Winterberry.**

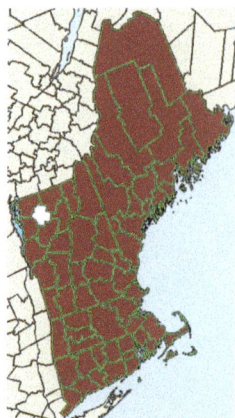

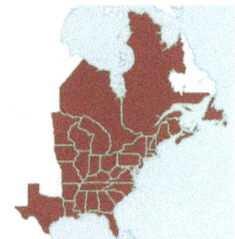

SKELICO

COMMON WINTERBERRY (*Ilex verticillata*)

BLAKE ROSS

MARK CHANDLER

■ Speckled Alder

Alnus incana (L.) Moench

GROWTH FORM Spreading or loosely ascending shrub to 8 m high, the brown or blackish-gray bark marked with whitish linear lenticels ("speckles") 7 mm or more long; pith of twigs triangular in cross-section.

LEAVES Deciduous, alternate, simple, oval, ovate, rounded to subcordate at bases, mostly doubly serrate or serrate-dentate, often with a wavy margin, rugose, 5–10 cm long.

FLOWERS In preformed catkins; the male flowers purplish, conspicuous, with 4 stamens; the female flowers smaller, developing into woody, cone-like structures in fruit, March–early June.

FRUIT A woody strobile or cone-like structure about 1.5 cm long, borne in inflorescences of 4–8 on short-stalked or sessile pedicels, autumn.

WHERE FOUND A common shrub along the bog margins, often persisting in the sphagnous mat; larger and more abundant on low ground, in swamps, and along the margins of water courses.

NOTES Often forming dense thickets in wet meadows, etc. where few other plants except willows (*Salix* spp.) would thrive, thus forming favorable habitat for wildlife, including beaver, which use the stems for lodges and dams.

OTHER NAMES *Aulne Blancharte* (Quebec), **Grey Alder**, **Hoary Alder**, *Verne* (Quebec). Formerly classified as *Alnus rugosa* (Du Roi) Spreng.

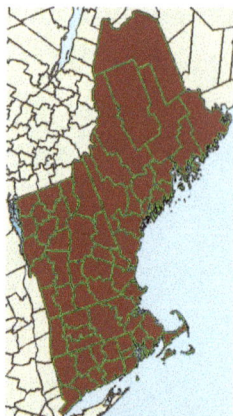

SPECKLED ALDER (*Alnus incana*)

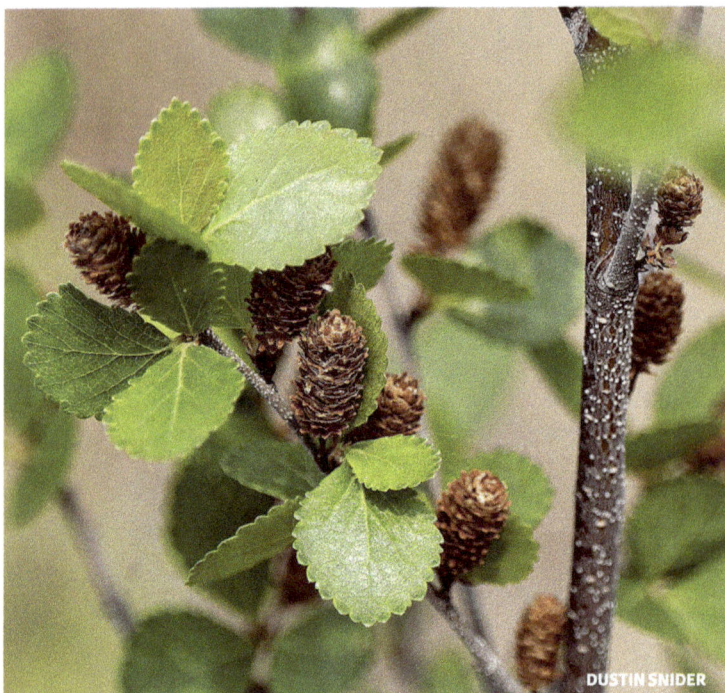

DUSTIN SNIDER

■ Bog Birch
Betula pumila L.

GROWTH FORM Erect or prostrate and matted, 0.5–5 m high, with glabrous to densely pubescent or tomentose, glandless branchlets.

LEAVES Deciduous, alternate, simple, coarsely or dentate-serrate, suborbicular to broad-elliptic or obovate, reticulate-veined, 0.8–7 cm long.

FLOWERS Staminate catkins elongated, formed in autumn and remaining naked during the winter; pistillate catkins (photo, above) terminating short lateral branches (spur shoots) of the season, May–June.

FRUIT A minute nutlet or samara, borne in sessile, erect, fragile strobiles 0.7–3 cm long and 5–9 mm thick, maturing in the fall.

WHERE FOUND An uncommon shrub of sphagnum bogs, growing better in calcium-rich fens and wooded swamps.

OTHER NAMES *Bouleau* (Quebec), **Low Birch**, **Swamp Birch**.

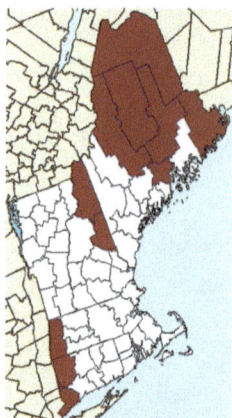

BONNIE SEMMLING

BOG BIRCH (*Betula pumila*)

REUVEN MARTIN

■ Swamp Fly-Honeysuckle

Lonicera oblongifolia (Goldie) Hook.

GROWTH FORM Erect shrub to 1.5 m high, with minutely pubescent, ascending branchlets.

LEAVES Deciduous, opposite, simple, entire, oblong to narrowly obovate, short-pubescent on both sides, bluish-green above and grayish green below, 3–8 cm long.

FLOWERS Corolla yellowish-white, deeply 2-lipped, borne on thread-like peduncles, 1.5–4 cm long, late May–June.

FRUIT An orange-yellow to deep red, subglobose berry, borne singly or united in pairs with few to many seeds, July, August.

WHERE FOUND Places in the bog, often near **Northern White-cedar** (*Thuja occidentalis*) where the area is underlain with limestone; also swampy thickets and wet woods.

NOTES Occasional, especially northward and in calcareous areas. The red fruits distinguish this species from the only other wetland honeysuckle, **Mountain-fly Honeysuckle** (*Lonicera villosa*).

OTHER NAMES *Chèvrefeuille* (Quebec).

ROB FOSTER

SWAMP FLY-HONEYSUCKLE (*Lonicera oblongifolia*)

DUSTIN SNIDER

■ Mountain Fly-Honeysuckle

Lonicera villosa (Michx.) J.A. Schultes

GROWTH FORM Low, depressed ascending or upright shrub, 1 meter or less tall, with glabrous to more or less pubescent or tomentose branches (depending on the variety), and shreddy, tan-brown bark.

LEAVES Deciduous, opposite, simple, oblong to oblong-lanceolate, entire, blunt-pointed, short-petioled, glabrous or pubescent on both sides, (pubescence varying with the variety), 2–8 cm long.

FLOWERS Yellow with subequal lobes, borne on short peduncles 1–7 mm long, late April–July.

FRUIT An ellipsoidal or subglobose, dark blue, bloomy, edible, several-seeded berry (the two ovaries surrounded by a blue fleshy cup, a portion of the ovaries still evident at the tips of the fruit), May–August.

WHERE FOUND More common near the borders of the bog, often in alder thickets; also in peaty or rocky barrens and grassy swales.

NOTES Often visited by bees, especially since the blossoms emerge before many other plants are in bloom. Several varieties are described by botanists.

OTHER NAMES Blue Fly-honeysuckle, *Chèvrefeuille* (Quebec), **Waterberry**.

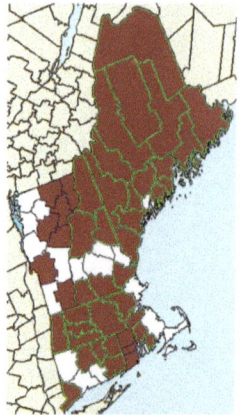

TOM NORTON

MOUNTAIN FLY-HONEYSUCKLE (*Lonicera villosa*)

GEOSESARMA

■ Atlantic White-Cedar

Chamaecyparis thyoides (L.) B.S.P.

GROWTH FORM Strong-scented evergreen conifer to 25 m high (often dwarfed in bogs), with upright, horizontal spreading, irregularly arranged branches forming a narrow spire-like head; bark of trunk shreddy, soft, reddish brown.

LEAVES Evergreen, simple, minute; scale-like and opposite, completely clothing the twigs and young branchlets (which are slightly flattened but not disposed in horizontal sprays, as in *Thuja*).

FLOWERS Monoecious, tiny, terminal, cone-like: the male flowers yellow, the female flowers purplish brown, March or April.

FRUIT Cones maturing in one growing season, subglobose, bluish purple, bloomy, 6–9 mm in diameter, not opening until maturity in the fall; cone-scales thick, peltate with terminal boss, wedge-shaped, attached at a central point (as opposed to the elongate cone-axis in *Thuja*). Occasional in sphagnum but more often in coastal swamps where it becomes commercially important southward.

NOTES The scented wood is exceedingly durable; used for the same purposes as Northern White-cedar.

OTHER NAMES Coast White-cedar, False Cypress, Swamp-cedar, Southern White-cedar, White-cedar.

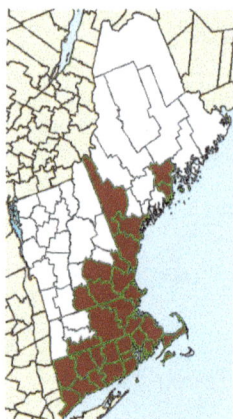

ATLANTIC WHITE-CEDAR (*Chamaecyparis thyoides*)

PUFFERCHUNG

■ Eastern Red-Cedar

Juniperus virginiana L.

GROWTH FORM Columnar or spire-like evergreen conifer to 30 m tall, with slender 4-angled or terete branchlets clothed with small scaly or awl-shaped leaves.

LEAVES evergreen, simple, small, dimorphic, either scale-like, opposite and tightly appressed on old growth, or awl-shaped, ternate, and spreading (on young plants or vigorous shoots).

FLOWERS Dioecious (rarely monoecious), borne laterally in small, scaly catkins or cone-like structures: the male flowers yellow, the female flowers bluish, May.

FRUIT A blue-black, berry-like cone 5–6 mm in diameter, with sweetish flesh, composed of 3–6 fused, fleshy scales, borne on straight pedicels, maturing in one growing season.

WHERE FOUND Occasional in the more open places in the bog or along the wet margins; more common and of better growth in dry, open woods, or on rocky slopes, and sandy or gravelly barrens, in neutral or acid soils.

NOTES Thrives on a variety of habitats from wet to dry. The beautiful, characteristically scented, durable wood is valuable for chests and cabinets. Trees are often infected with unsightly cedar-rust galls, formed by a fungus that also infects nearby apple trees.

OTHER NAMES *Cèdre Rouge* (Quebec), **Northern Red-cedar, Red-cedar, Red Juniper, Savin.**

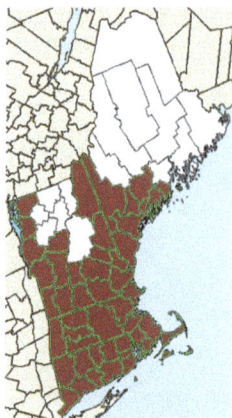

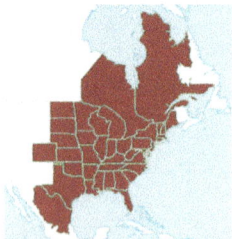

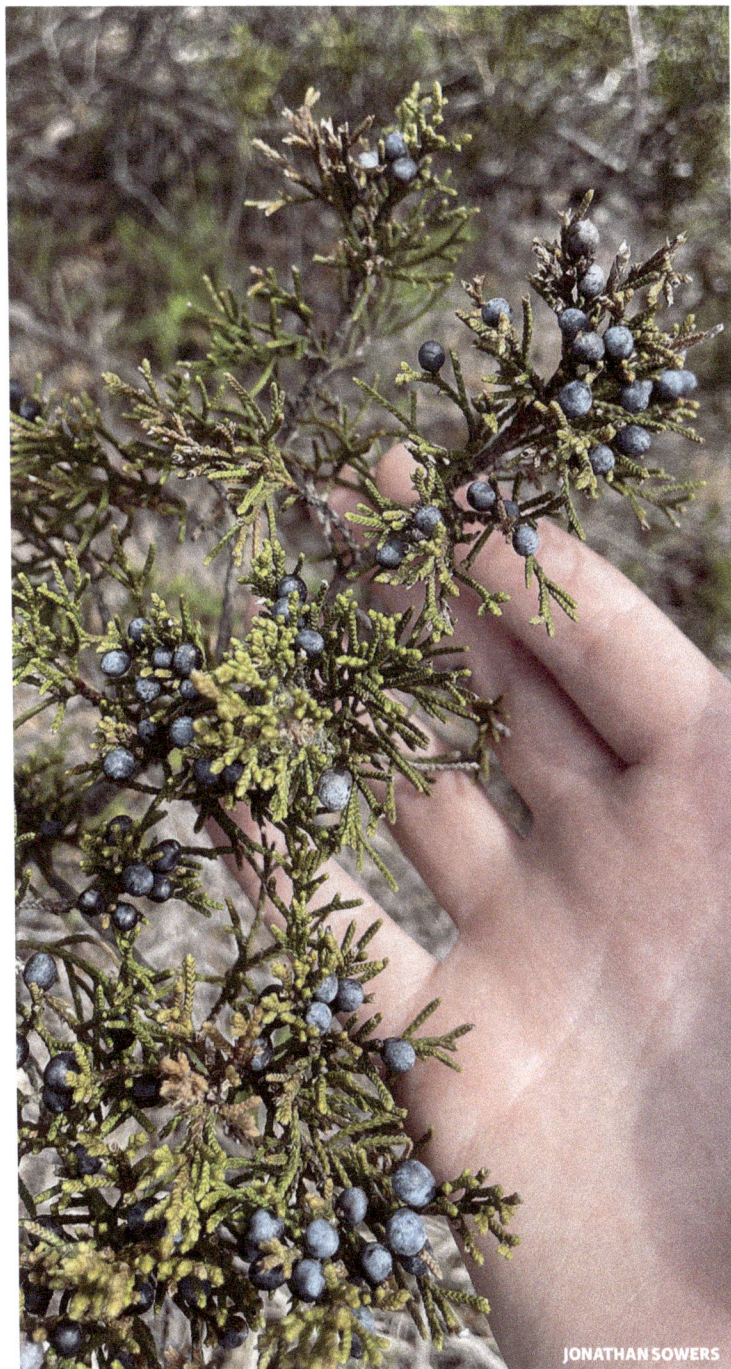

JONATHAN SOWERS

NORTHERN RED-CEDAR (*Juniperus virginiana*)

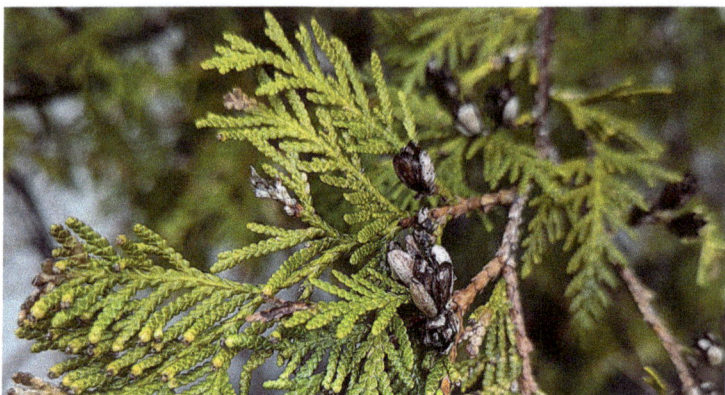

■ Northern White-Cedar
Thuja occidentalis L.

GROWTH FORM Pyramidal, strong-scented, evergreen conifer to 20 m high (smaller and stunted in bogs) with short, spreading branches, strongly flattened branchlets (all in one plane) clothed with small, scale-like, overlapping leaves; trunks buttressed, with shreddy, red-brown, soft bark.

LEAVES Evergreen, simple, small, scale-like and opposite, overlapping in 4 rows, glandular on the back, completely clothing the twigs and young branchlets which are disposed in flat, horizontal sprays.

FLOWERS Monoecious, tiny, cone-like, composed of few scales: male flowers yellow; female flowers purplish or reddish, April or May.

FRUIT Cones erect, oblong or elliptic-oblong, light or yellowish brown, 8 mm long, with 8–10 thin, flat scales (only 4 fertile), attached at different points to an elongate axis, maturing in one growing season and remaining on the branchlets for about a year.

WHERE FOUND Often getting a "foothold" here and there in the deep sphagnum where it is of poor, stunted growth; developing better near the border of the bog where mineral matter is available; attaining better development in swamps and on cool rocky banks, especially where the soil is limy.

NOTES Wood scented, light, soft, and durable; often used for poles, posts, railroad ties, etc.; important to wildlife as a browse and for shelter.

OTHER NAMES American Arborvitae, *Balai* (Quebec), *Cèdre* (Quebec), Eastern Arborvitae, Eastern White-cedar, Oo-soo-ha-tah ("Feather-leaf"), Swamp-cedar.

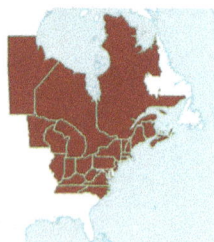

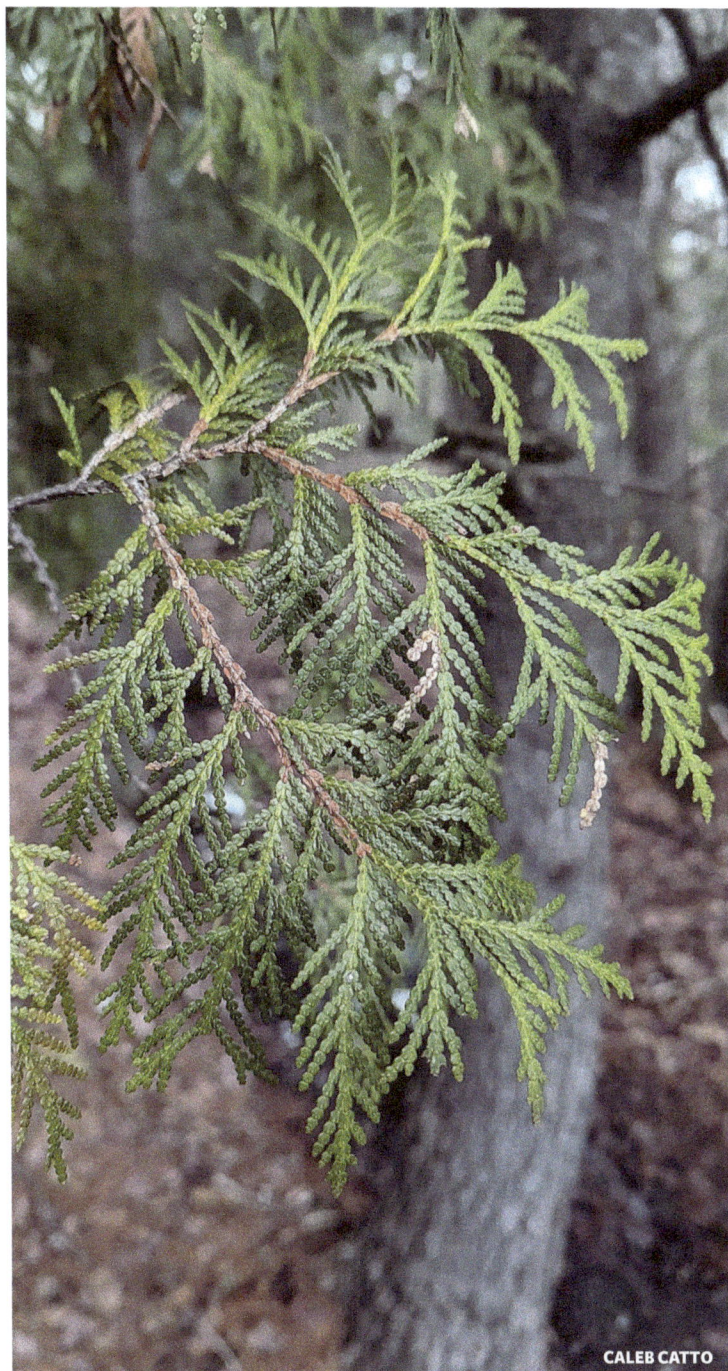

CALEB CATTO

NORTHERN WHITE-CEDAR (*Thuja occidentalis*)

DEMIAN HISS

■ Bog-Rosemary

Andromeda polifolia L.

GROWTH FORM Low, pale, evergreen shrub with elongate creeping base and ascending, nearly terete, little-branched, glaucous stems 1–7 dm high.

LEAVES Evergreen, alternate, simple, firm, revolute, linear to narrowly oblong, white-tomentulose beneath.

FLOWERS White, terminal, nodding, May–July.

FRUIT A depressed-globose, glaucous capsule with rather thick valves, borne on short, thick, recurved pedicels.

WHERE FOUND Abundant, its creeping root-stalks composing a considerable portion of the fibrous mat of peat bogs, able to thrive where many other woody plants could not survive in the thick accumulation of living sphagnum; also common in peat, on margins of pools and other continuously wet places where it may not be associated with sphagnum.

OTHER NAME Andromeda, Wild Rosemary.

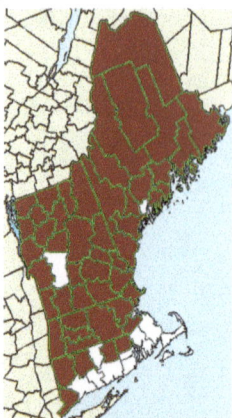

BOG-ROSEMARY (*Andromeda polifolia*)

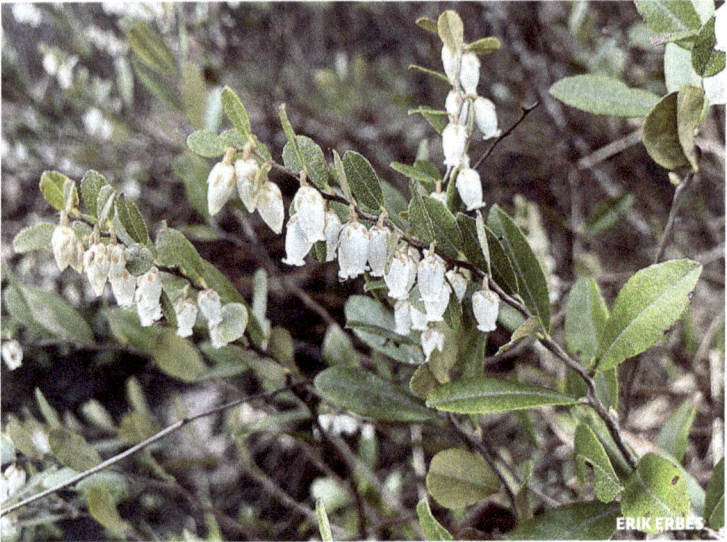

ERIK ERBES

■ Leatherleaf

Chamaedaphne calyculata (L.) Moench

GROWTH FORM Upright, low, much-branched, nearly evergreen, circumpolar shrub to 1.5 m tall, with scaly buds and slender twigs; branchlets slender, spreading or horizontal, tan-brown, appearing white-striped by numerous, linear or narrowly elliptic white lines due to exfoliating of the thin, outer, waxy or scurfy layer.

LEAVES Nearly evergreen, alternate, simple, entire (often with wavy margins), revolute, oblong to oblong-lanceolate, 2.5–5 cm long and one-third as wide (becoming progressively smaller toward the tip of the twig), dull green and slightly scaly above, densely scurfy below.

FLOWERS Axillary; corolla white, ureolate, 6–7 mm long, March–July.

FRUIT Capsules depressed, many-seeded, 4 mm across, tipped by a persistent style (resembling a pin stuck in a pin cushion).

WHERE FOUND Abundant in bogs, especially near the border where the sphagnum is less dense; also common in peaty swales, pond margins, etc., where water is plentiful.

NOTES Eagerly visited by bees early in the spring because the flowers appear before those of most other plants, and bees take advantage of the early treat.

OTHER NAME Cassandra.

LAURA J. COSTELLO

LEATHERLEAF (*Chamaedaphne calyculata*)

SIGITAS JUZÉNAS

KATHERINE BAIRD

■ Black Crowberry

Empetrum nigrum L.

GROWTH FORM Evergreen, procumbent and spreading shrub with creeping branches, partially buried and glabrous or minutely glandular, very slender branchlets, sometimes forming large mats when growing in humus.

LEAVES Evergreen, alternate, entire, linear to narrowly elliptic or linear-oblong, crowded, glabrous or glandular pulverulent, deeply furrowed beneath, reflexed and divergent at maturity, 2.5–7 mm long.

FLOWERS Dioecious or monoecious, axillary, inconspicuous with 3 petal-like purplish sepals and 3 pinkish stamens, flowering in June or late July.

FRUIT A berry-like black (rarely purple or whitish) drupe with 6–9 seedlike nutlets, and watery pulp, July–November.

WHERE FOUND Occasionally in peaty bogs but more common in peaty soil along the coast or in cool alpine areas.

NOTES An ornamental shrub occasionally used in rock-gardens where it forms dense, evergreen patches.

OTHER NAMES *Corbigeau* (Quebec), **Curlewberry**, *Graines a Corbigeau* (Quebec).

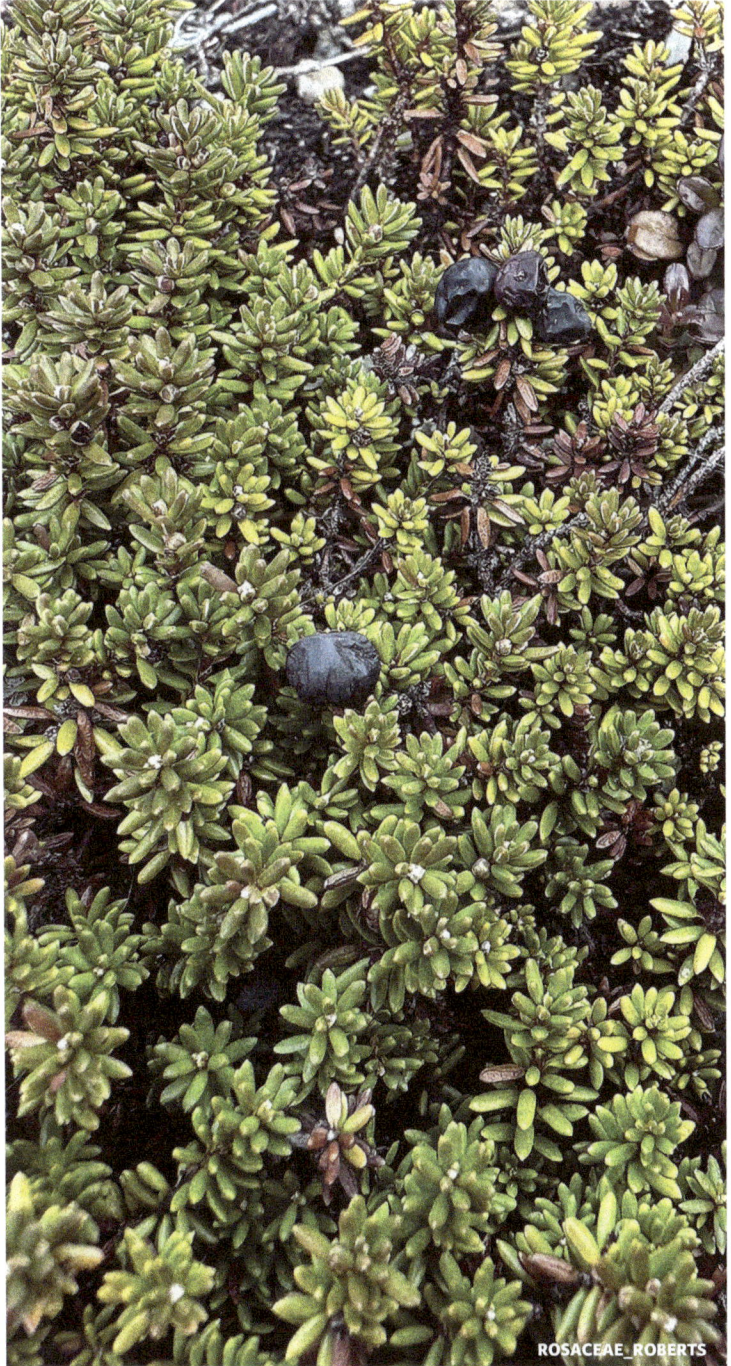

ROSACEAE_ROBERTS

BLACK CROWBERRY (*Empetrum nigrum*)

ERIK ERBES

■ Creeping Snowberry
Gaultheria hispidula (L.) Muhl. ex Bigelow

GROWTH FORM Trailing and creeping, delicate, matted evergreen plant with slightly woody stems and coarsely appressed setose, yellowish-brown or tan colored branchlets.

LEAVES Evergreen, alternate, simple, entire, orbic-ular-ovate to ovate, narrowed at the ends, rev-olute, lustrous and glabrous above, paler beneath and rusty-strigose on midrib, 4–10 mm long.

FLOWERS 4 mm long, white, mostly solitary in the axils on short, nodding peduncles, April or May (or until August in the mountains).

FRUIT A shiny, white, subglobose, many-seeded capsule (enclosed when ripe by a fleshy calyx so as to appear as a globular, slightly bristly, juicy, slightly acid, aromatic (wintergreen-tast-ing) berry), 6 mm in diameter.

WHERE FOUND Forming dense, bright green car-pets in the bog, studded with white "berries" in summer; also in mossy coniferous woods and uplands.

NOTES The leaves and fruits have a mild winter-green flavor and have been used as a food and medicine.

OTHER NAMES Capillaire, Maidenhair-berry, Moxie-plum, *Oeufs de Perdrix* (Quebec), *Petit Thé* (Quebec).

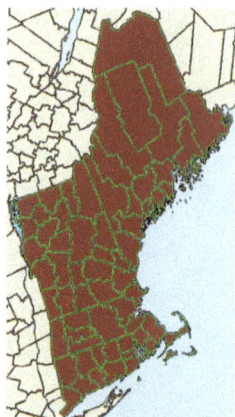

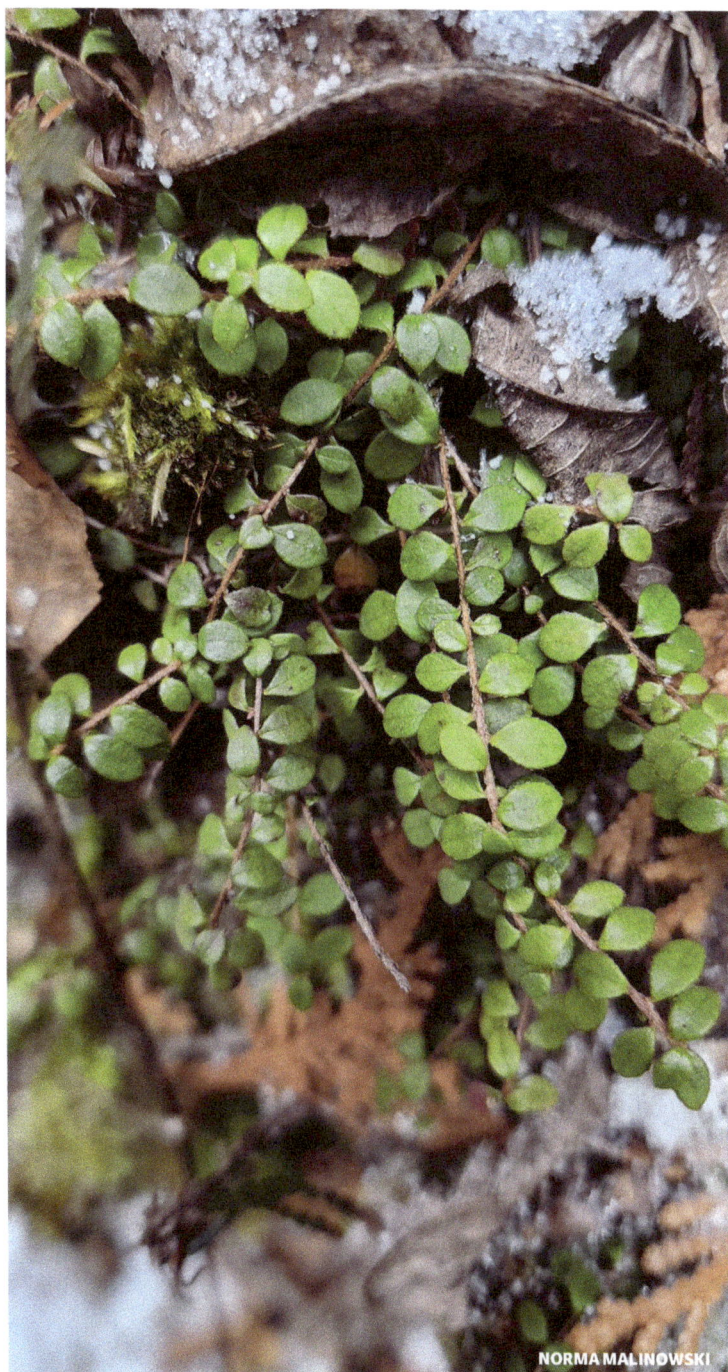

NORMA MALINOWSKI

CREEPING SNOWBERRY (*Gaultheria hispidula*)

ZIHAO WANG

■ Black Huckleberry

Gaylussacia baccata (Wangenh.) K. Koch

GROWTH FORM Stiff, upright, much-branched, slightly pubescent shrub, 0.3–1 m high, with densely resinous and sticky young growth.

LEAVES Deciduous. alternate, simple, entire, elliptic to oblonglanceolate or elliptic-obovate, abundantly resinous-dotted on both sides, 2.5–5 cm long.

FLOWERS Corolla conic-ovoid, dull red, 5 mm long, borne on pedicels 2–8 mm long, covered with shining, resinous globules, May–July.

FRUIT Berry-like, lustrous, black drupes, 6–8 mm across, borne on pedicels 2–8 mm long, August to October.

WHERE FOUND Moist, open places in the bog; also dry or moist woods and clearings elsewhere.

NOTES Fruit sweet and edible except for the rather large, nutlike seeds.

OTHER NAME *Gueules noires* (Quebec).

STEVEN LAMONDE

BLACK HUCKLEBERRY (*Gaylussacia baccata*)

DAVID MCCORQUODALE

■ Dwarf Huckleberry

Gaylussacia bigeloviana (Fern.) Sorrie & Weakley

GROWTH FORM Low, slender shrub 1–5 dm high from a subterranean base, with erect, somewhat glandular-hairy branches.

LEAVES Deciduous, alternate, simple, entire, mucronate, cuneateoblanceolate to oblong-obovate, persistently glandular-pubescent on both surfaces, 2.5–4 cm long.

FLOWERS Bell-shaped, white, pink, or red, 8–9 mm long, borne in loose, persistent, leafy-bracted racemes, June–August.

FRUIT Berry-like, black, glandular-pubescent drupes, 8–12 mm across, with 10 one-seeded nutlets, August to October.

WHERE FOUND Sphagnous bogs, often on large, floating mats; also on wet peats.

NOTES Fruit insipid. Much less common than the **Black Huckleberry** (*Gaylussacia baccata*).

OTHER NAME Gopherberry.

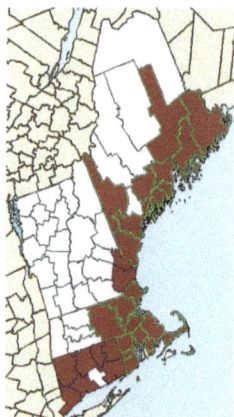

IAN MANNING

DWARF HUCKLEBERRY (*Gaylussacia bigeloviana*)

ALINA MARTIN

■ Sheep-Laurel

Kalmia angustifolia L.

GROWTH FORM Slender evergreen shrub up to 1.7 m high. Branchlets terete, strongly ascending, glabrous or nearly so, with naked buds.

LEAVES Evergreen, simple, opposite or ternate, entire, flat, rather thin, oblong to elliptic-lanceolate, glabrous or puberulent and glabrate, (ferruginous when young), on short petioles, obtuse or acutish, bright green above, paler beneath; blades often bending down due to the curved petioles.

FLOWERS Deep rose-pink or purplish crimson (rarely white), borne laterally on slender pedicels along the stem; pedicels glandularpuberulent, recurving in fruit, blooming in late May–August.

FRUIT A rather small, thin-valved, fragile, depressed-globose capsule.

WHERE FOUND A small shrub commonly found on wet sterile soil, old pastures and barrens (often a weed in blueberry fields), but commonly invading and surviving in sphagnous bogs.

NOTES Leaves poisonous to livestock.

OTHER NAMES Dwarf Laurel, Pig-laurel, Lambkill, Wicky.

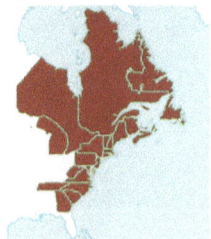

BRENDAN BOYD

SHEEP-LAUREL (*Kalmia angustifolia*)

DAVID MCCORQUODALE

■ Bog-Laurel

Kalmia polifolia Wangenh.

GROWTH FORM Slender straggling shrub to 0.7 m tall. Branchlets, glabrous or puberulous, 2-edged (i.e. with flattened, winglike margins).

LEAVES Evergreen, opposite (rarely in 3's), simple, sessile or nearly so, 0.7–3.5 cm long, oblong to lanceolate or linear, often revolute, entire, firm with blunt callous tip, glabrous, lustrous-green above, conspicuously whitened beneath, with prominent midrib.

FLOWERS Terminal, deep pink to crimson or rose-purple, blooming in mid-May to July.

FRUIT A globose-ovoid, rather thin-valved, promptly-dehiscing capsule, borne on a long, slender, erect pedicel.

WHERE FOUND One of the common plants forming a considerable portion of the fibrous mat of the bog.

NOTES The flowers add local color to the bog in the spring.

OTHER NAMES Pale Laurel, Swamp Laurel.

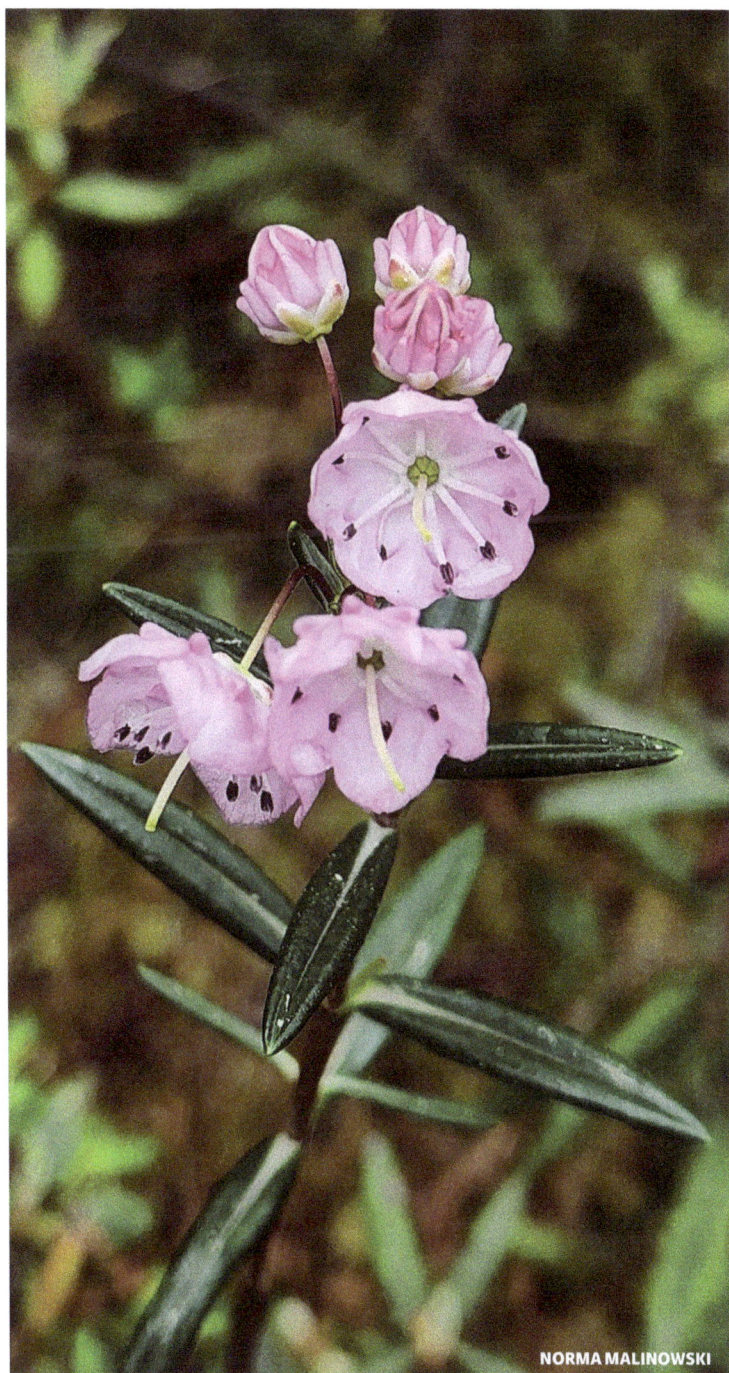

NORMA MALINOWSKI

BOG-LAUREL (*Kalmia polifolia*)

STEVEN LAMONDE

■ Maleberry

Lyonia ligustrina (L.) DC.

GROWTH FORM Glabrous, or somewhat pubescent, much-branched shrub to 4 m tall, with tan-brown twigs and ashy gray branchlets; outer bark thin, shreddy.

LEAVES Deciduous, alternate, simple, serrulate to entire, lanceolate or oblanceolate to ovate or broadly elliptic, sometimes with scurfy scales, 3–7 cm long.

FLOWERS Corollas whitish, globose- or ovoid-urn-shaped, 2.5–5 mm long, May–July.

FRUIT A 5-angled, globose or subglobose capsule, 2–5 mm long and 3 mm across, with thickened sutures.

WHERE FOUND Open places or near the margins of the bog where the sphagnum is not dense; also in wet or dry thickets.

NOTES Often mistaken for **Highbush Blueberry** (*Vaccinum* spp.), especially when in flower. Foliage poisonous to young stock.

OTHER NAME Male Blueberry, He-Huckleberry.

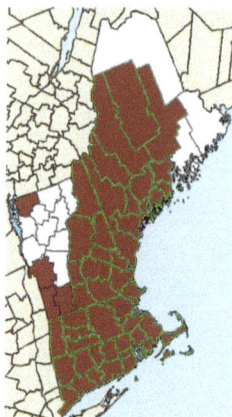

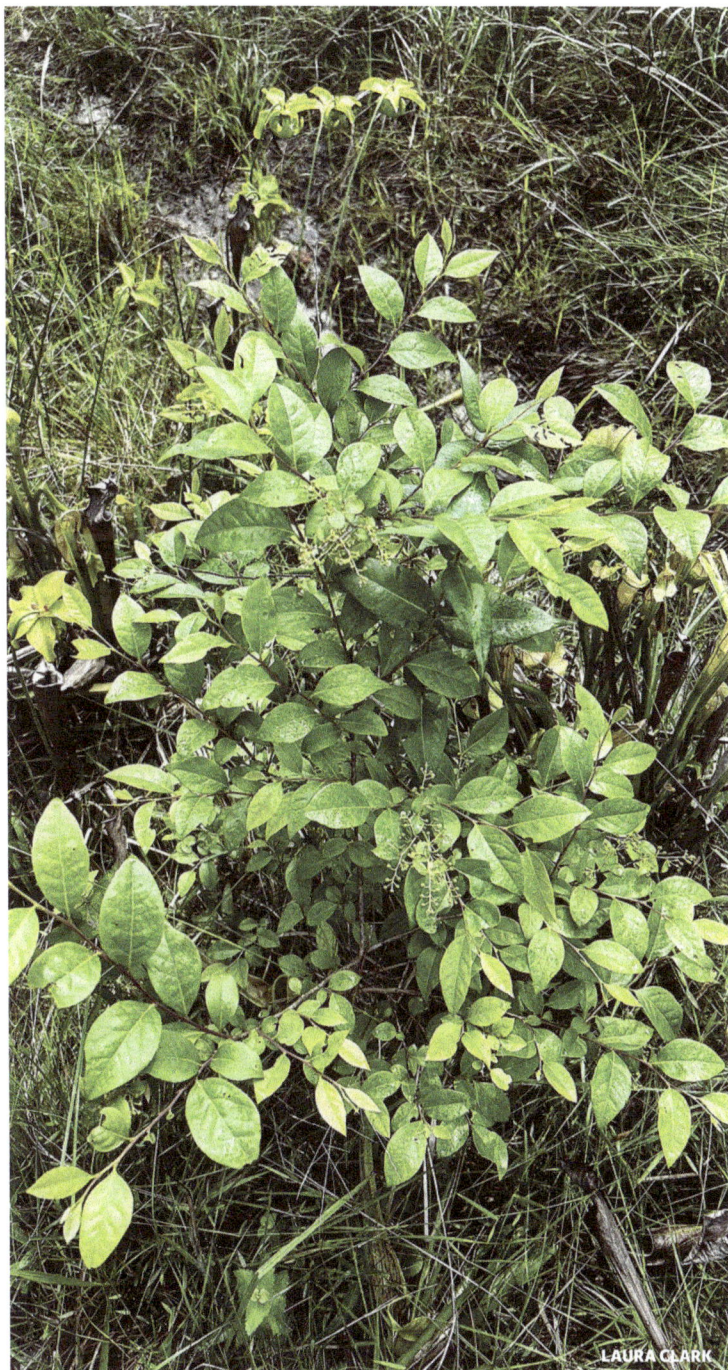

LAURA CLARK

MALEBERRY (*Lyonia ligustrina*)

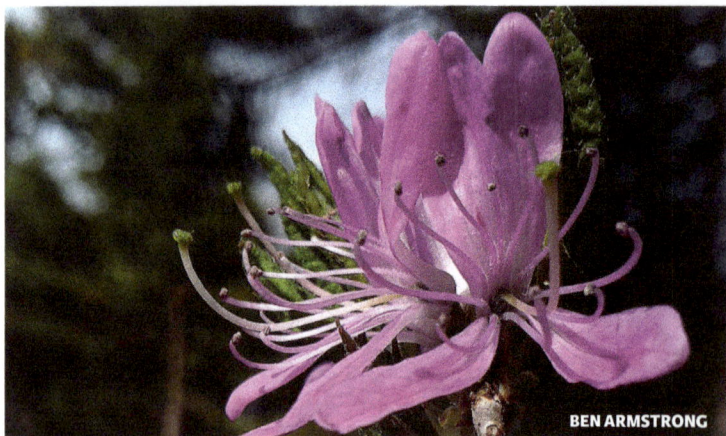

BEN ARMSTRONG

■ Rhodora

Rhododendron canadense (L.) Torr.

GROWTH FORM Low, profusely branched shrub usually less than 1 meter high with strongly ascending branches. Branchlets puberlous when young, yellowish red to pinkish, bloomy.

LEAVES Deciduous, alternate, simple, petioled, elliptic to oblong, obtuse to acute at tip, cuneate at base, 2–6 cm long; margins entire, somewhat revolute, ciliate; dull grayish or bluish green, shining and glaucous or pubescent above; sparsely grayish tomentulose, usually sparingly glandular and pilose on the midrib beneath.

FLOWERS Terminal, rose purple (or rarely white), conspicuous due to the expanding of the inflorescences before the leaves, blooming March–July.

FRUIT A slender persistent, glaucous-puberlent capsule 0.7–1.5 cm long, tipped by a long, slender style.

WHERE FOUND Common in the more open areas where sphagnum is sparse. Locally abundant in damp thickets, acid barrens and rocky slopes as well as in sphagnous bogs.

NOTES When in flower in the spring, one of the most conspicuous and beautiful plants of the bog. Bees sometimes become tangled (hobbled) while visiting the flowers and come in contact with the sticky cobwebby threads which cling to the pollen grains as they are released from the anthers. Beekeepers lose part of their colonies each year due to this phenomenon known as "bee tanglefoot".

OTHER NAMES None.

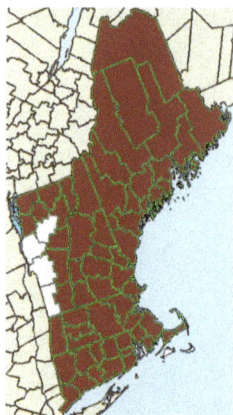

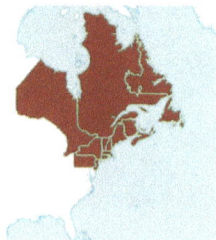

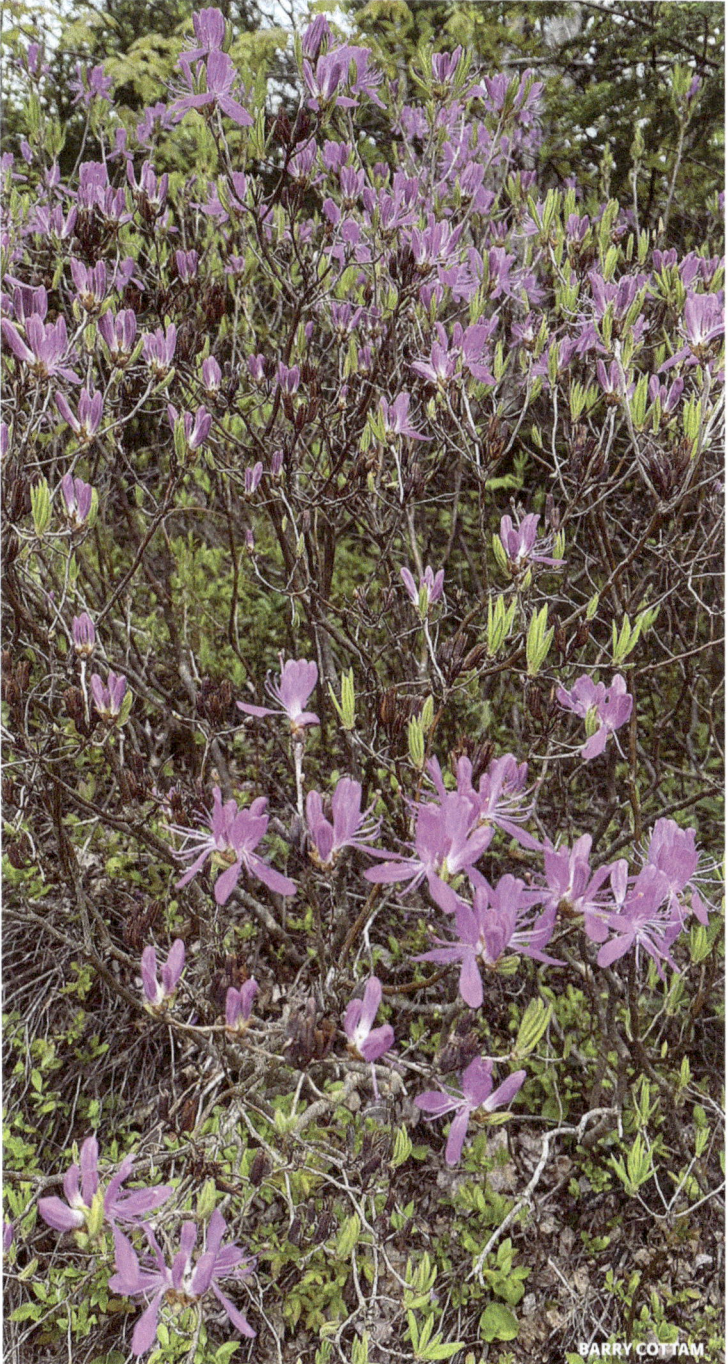

BARRY COTTAM

RHODORA (*Rhododendron canadense*)

TOM SCAVO

■ Labrador Tea

Rhododendron groenlandicum (Oeder) K.A. Kron & Judd

GROWTH FORM Low boreal shrub, 1 meter or less high, rooting freely in the sphagnum and peat.

LEAVES Evergreen, alternate, simple, entire, with revolute margins, densely covered with white or rusty, woolly hairs beneath, oblong, or linear-oblong, obtuse, 2–5 cm long, their blades often directed toward the base of the twig due to a bend in the petiole. Bruised foliage and other succulent parts yielding a strong, resinous, fragrant aroma.

FLOWERS Small, white, borne in terminal, umbel-like clusters; corolla of 5 obovate, spreading petals, blooming in May or June.

FRUIT A slender capsule with 5 locules, splitting from the base upward, containing many tiny seeds. The capsules, splitting from the base, resemble a partially open umbrella or a shooting star. This method of dehiscence is unique and not likely to be encountered in any other local bog plants. (Capsules of other plants split open at tips or along the sides to liberate the seeds.)

WHERE FOUND Common, especially along margins of the bog where it is associated with speckled alder and other tall woody plants.

NOTES The leaves have been used for tea.

OTHER NAMES Formerly classified as *Ledum groenlandicum* Oeder.

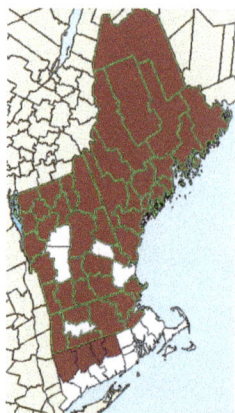

SIMONE L

LABRADOR TEA (*Rhododendron groenlandicum*)

ALINA MARTIN

■ Low Sweet Blueberry

Vaccinium angustifolium Ait.

GROWTH FORM Small, clone-forming, much-branched shrub up to 6 dm high, with glabrous or somewhat pilose, minutely warty ("goose-pimply") new branchlets.

LEAVES Deciduous, alternate, simple, lanceolate to oblong, 1.5–3.5 cm long and 8–15 mm wide, glabrous on both sides (or only slightly pilose on the midrib beneath, with closely and finely spinulose-serrulate margins, exhibiting various shades of green, blue-green, and red throughout the season.

FLOWERS Corolla cylindric-campanulate, white or tinged with pink, 6–10 mm long, early spring.

FRUIT A bloomy, blue or bluish-black, sweet delicious, edible berry, 6–15 mm in diameter, June–September.

WHERE FOUND Occasional in thin sphagnum and in mineral soil along the bog borders, but abundant in dry, open barrens (where it is commercially important), peats and rocks, extending to high elevations in the mountains.

NOTES Thhe principal commercially important lowbush species. The flowers of all species of blueberries are eagerly visited by bees (upon which the blueberry growers depend for pollination and an ultimate successful harvest of fruits), for pollen and nectar.

OTHER NAMES *Bleuet* or *Bluet* (Quebec), **Lowbush Blueberry, Late Sweet Blueberry, Sweet Hurts.**

JONATHAN SOWERS

LOW SWEET BLUEBERRY (*Vaccinium angustifolium*)

ER-BIRDS

■ Highbush Blueberry
Vaccinium corymbosum L.

GROWTH FORM Shrub to 4 m tall, with stiff, spreading, somewhat gnarled or distorted branches, forming compact or open clumps; young branchlets yellow-green, minutely warty ("goose-pimply").

LEAVES Deciduous, alternate, simple, ovate to elliptic-lanceolate, entire (or sometimes finely serrulate), glabrous, half-grown at flowering time, 4–8 cm long and 2–4 cm broad, the lower surface pubescent on the veins, green on both sides (or sometimes glaucous beneath).

FLOWERS In dense clusters, corolla white (sometimes pink-tinged), ovoid to cylindric-urceolate, 6–12 mm long and 4–6 mm wide, early spring.

FRUIT A blue-black, bloomy, many-seeded, edible (sweet and juicy) berry 6–12 mm in diameter, late June to early September.

WHERE FOUND Mostly in open places or along the wet borders of the bog; also in swamps, low woods, and occasionally in rocky or dry uplands, sparse or absent northward.

NOTES An interesting shrub because of its delicious fruits, and foliage (which turns bright scarlet and orange in the fall). Also, as in other species of blueberries, much visited by bees for nectar and pollen. Unsightly "witches' brooms" are often present on the branches.

OTHER NAMES *Bleuet* or *Bluet* (Quebec), **Swamp Blueberry**.

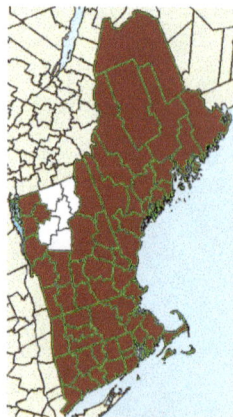

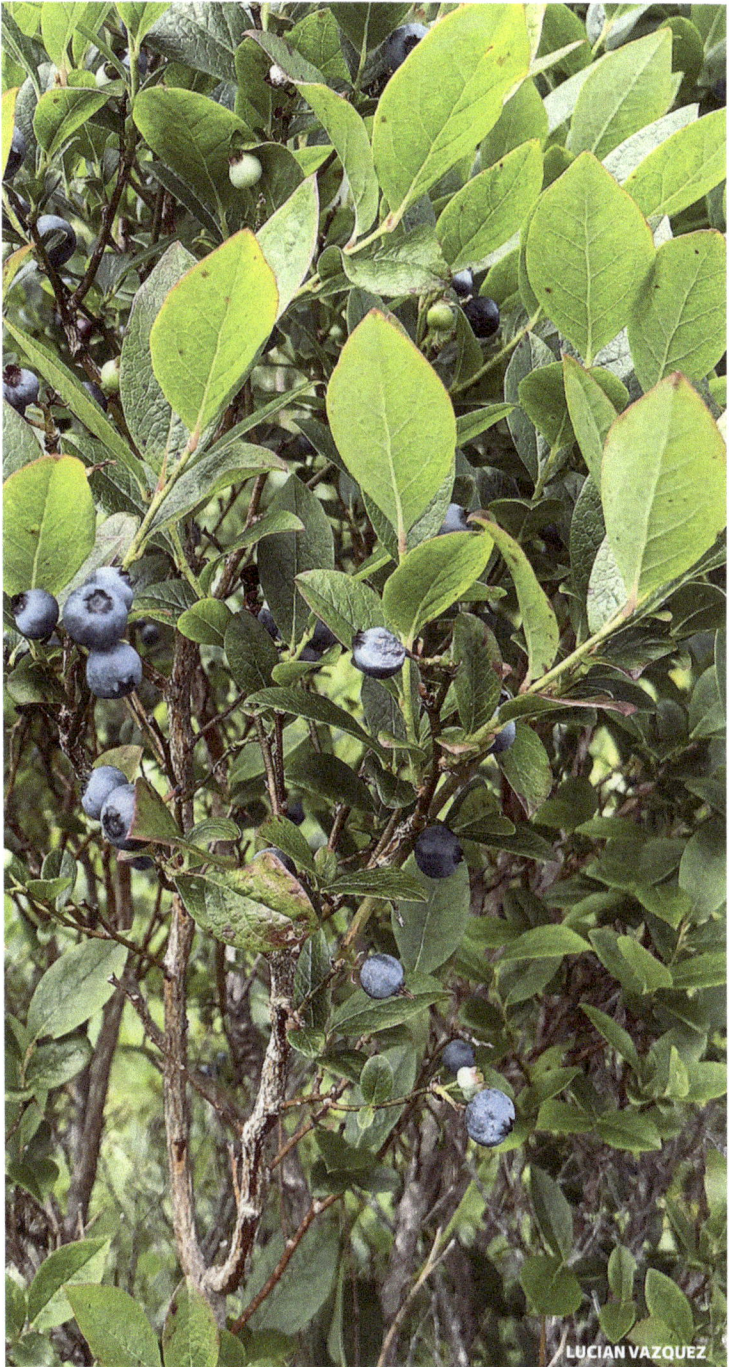

LUCIAN VAZQUEZ

HIGHBUSH BLUEBERRY (*Vaccinium corymbosum*)

BOBBY MCCABE

BOBBY MCCABE

■ Black Highbush Blueberry

Vaccinium fuscatum Ait.

Similar to **Highbush Blueberry** (*Vaccinium corymbosum*) except in the following respects:

LEAVES Moderately to densely hairy on underside, entire, unexpanded at flowering time.

FLOWERS Corolla ovoid to elliptic, yellowish or greenish-white tinged with purple, 5–8 mm long.

FRUIT Berries shiny black without a bloom, 5–8 mm in diameter, ripening a week or so earlier than that of *Vaccinium corymbosum*.

NOTES Flowers and fruits a week to ten days before *Vaccinium corymbosum*.

OTHER NAMES *Bleuet* or *Bluet* (Quebec), **Downy Swamp Blueberry**.

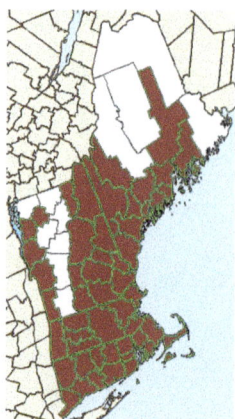

DAVID MCCORQUODALE

BLACK HIGHBUSH BLUEBERRY (*Vaccinium fuscatum*)

LOUIS IMBEAU

■ Large Cranberry
Vaccinium macrocarpon Ait.

GROWTH FORM Stems slender and creeping, elongate and intricately forking, the flowering branches ascending, rooting in the sphagnous mat.

LEAVES Evergreen, alternate, simple, oblong-elliptic, blunt at tip, 6–18 mm long, 2–8 mm broad, pale or slightly whitened beneath, flat or slightly revolute.

FLOWERS Corolla-segments 6–10 mm long, roseate; pedicels 1–10, arising laterally from an elongate rachis 1–3 cm long.

FRUIT Red, (often pale and speckled when immature), acid, globose, ellipsoid, obovoid or pyriform berry 1–2 cm in diameter, September–November, borne on a long, thread-like pedicel, holding through winter.

WHERE FOUND Open places in the bog where it roots in the sphagnum; also found in swamps on wet shores and in meadows.

NOTES Cultivated for its berries which are used extensively in making jellies and preserves.

OTHER NAMES *Atocas* (Quebec), **American Cranberry.**

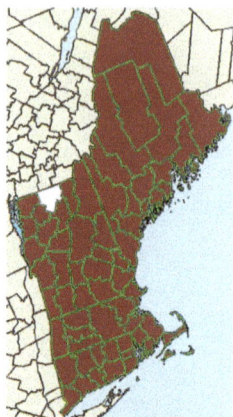

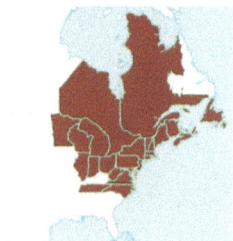

ERIK ERBES

LARGE CRANBERRY (*Vaccinium macrocarpon*)

ERIK ERBES

■ Velvet-leaf Blueberry

Vaccinium myrtilloides Michx.

GROWTH FORM Low, much-branched, clone-forming plant, 2–9 dm high with minute, warty ("goose-pimply"), fine, wiry, densely velvety branchlets.

LEAVES Deciduous, alternate, simple, narrowly elliptic to oblong-lanceolate, entire, downy beneath and often also above (appearing bluish-green or sage-like), 2–5 cm long and 0.5–2.5 cm broad, half grown at flowering time.

FLOWERS In dense, terminal clusters; corolla globose-urceolate or short-campanulate, 4–6 mm long, greenish or tinged with purple, early spring.

FRUIT An edible, blue (rarely whitish), rather sour berry 7–10 mm in diameter, with a heavy bloom, July–September.

WHERE FOUND Occasional in the bog but more abundant in moist, open woods, swamps and clearings, including dry, sandy barrens where it is harvested along with other commercial lowbush species.

NOTES Fruit less sweet and often smaller than those of other species; plants appearing more "leafy" than other species, thus partially obscuring the berries.

OTHER NAMES *Bleuet* or *Bluet* (Quebec), **Canada Blueberry, Sour-top Blueberry.**

NICK KLEINSCHMIDT

VELVET-LEAF BLUEBERRY (*Vaccinium myrtilloides*)

THIERRY ARBAULT

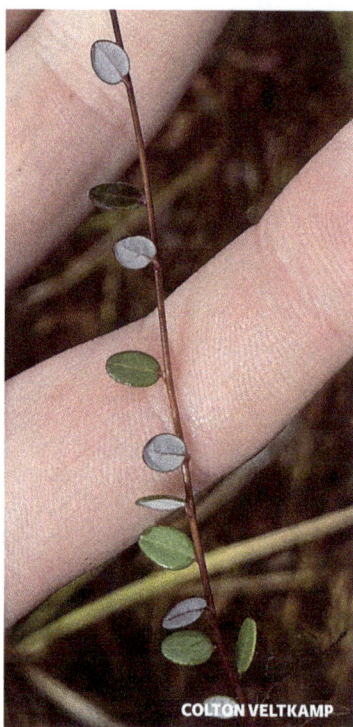

COLTON VELTKAMP

■ Small Cranberry

Vaccinium oxycoccus L.

GROWTH FORM Stems extremely slender and thread-like with ascending tips, the branchlets creeping and rooting in the sphagnum.

LEAVES Evergreen, alternate, simple, ovate-oblong to ovate or triangular, 3–8 mm long, 1–3 mm broad, strongly revolute, conspicuously whitened or glaucous beneath.

FLOWERS Corolla-segments roseate, 5–6 mm long, pedicels 1–4, arising from a terminal short rachis at most 3 or 4 mm long.

FRUIT Red, (often pale and speckled when young) acid, 5–8 mm in diameter, globose to pyriform, borne on a long, filamentous pedicel, holding over winter, August–October.

WHERE FOUND Open places in the bog, where it branches and roots freely; also peaty and upland soils.

NOTES Berries sour, used for jellies and preserves.

OTHER NAME *Atocas* (Quebec).

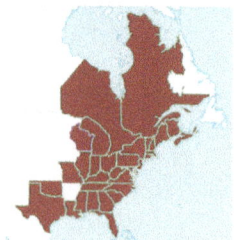

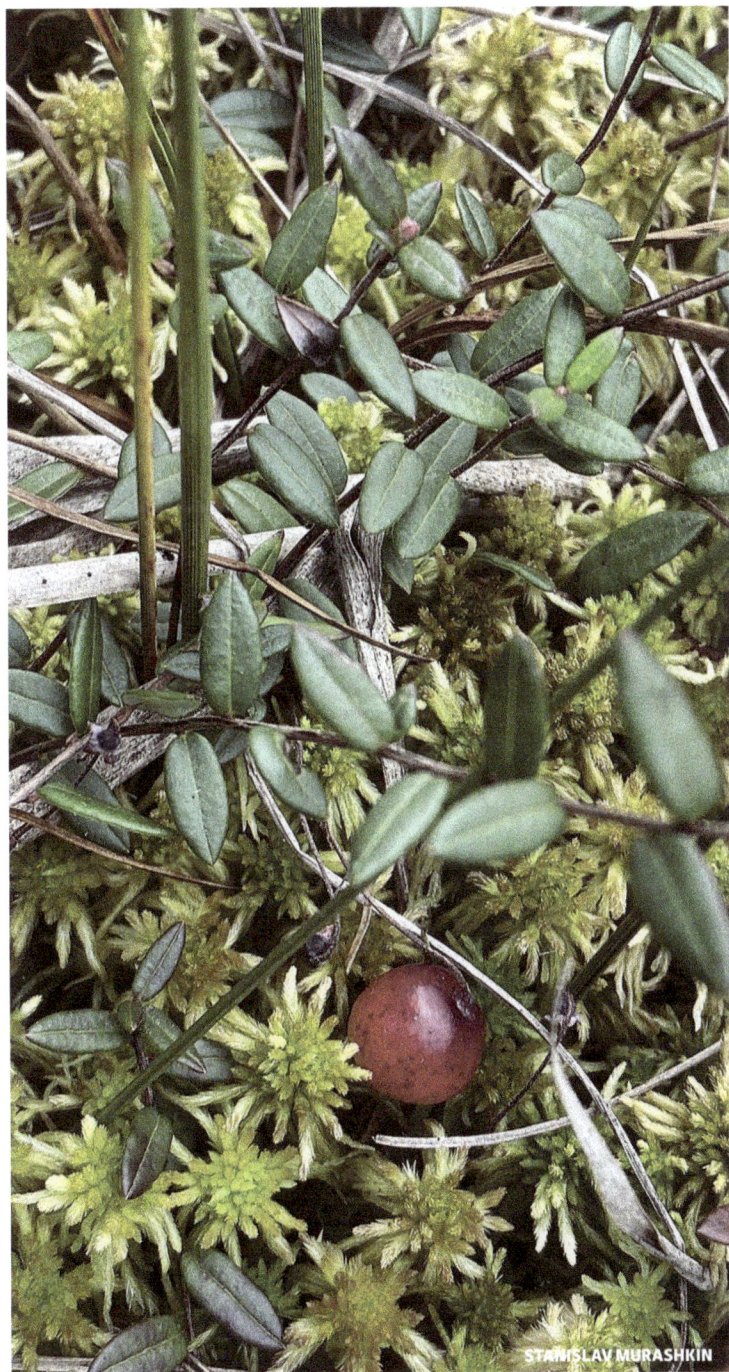

STANISLAV MURASHKIN

SMALL CRANBERRY (*Vaccinium oxycoccus*)

ER-BIRDS

■ Water-Willow
Decodon verticillatus (L.) Ell.

GROWTH FORM Aquatic smooth or downy sub-shrub with angled, recurved stems, 4–6 dm long, the arching stems rooting at the tips; bark of submersed stems conspicuously spongy-thickened.

LEAVES Deciduous, simple, opposite or whorled, lanceolate, nearly sessile, entire with undulate margins, 3–20 cm long.

FLOWERS Borne on short pedicels in clusters of 2–8 in the axils of the upper leaves; petals magenta or rose-purple, 8–12 mm long, July–August.

FRUIT A 3–5 valved, subglobose capsule 5 mm in diameter, bearing angular seeds.

WHERE FOUND Open places in the bog, growing in water; also swamps and shallow pools.

NOTES Occasional in bogs: unique because of its excessive development of soft spongy cork on the submersed parts of stems.

OTHER NAME Water Oleander.

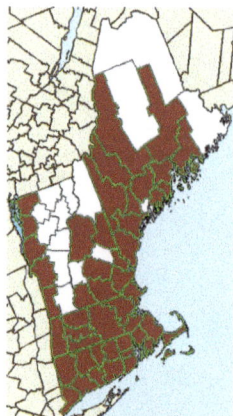

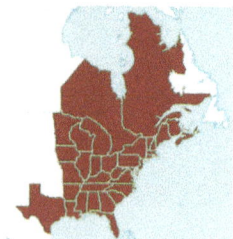

WATER-WILLOW (*Decodon verticillatus*)

ERIK ERBES

■ Sweet Gale
Myrica gale L.

GROWTH FORM Aromatic shrub to 2 m high with brown, wand-like, strongly ascending branches.

LEAVES Deciduous, alternate, simple, cuneate-oblanceolate, dark green and glabrous above, grayish-glabrous or more or less pubescent beneath, toothed toward the apex, 3–6 cm long.

FLOWERS Male catkins (above) 7–10 mm long; female catkins (right) cone-like, dense, 8–10 mm long, April–June.

FRUIT Dense, cone-like, 8–10 mm long; individual fruits ovoid, imbricated, 2-winged nutlets, resin-dotted, borne in firm, cone-like structures at the ends of the branchlets.

WHERE FOUND Open places in the bog or along the wet margins, often growing in water; also on wet shores and in swamps.

NOTES Scented shrub with pleasant odor. One of the most common woody plants on wet shores.

OTHER NAMES *Bois-sent-bon* (Quebec), "Meadow Fern", *Piment Royal* (Quebec).

THIERRY ARBAULT

SWEET GALE (*Myrica gale*)

DANIEL PATTERSON

■ Black Tupelo

Nyssa sylvatica Marsh.

GROWTH FORM Large tree with fissured bark and tough wood, to 30 m tall (much smaller in sphagnous bogs), with long, stiff, horizontal branches, usually forming a broad, flat-topped head; twigs with diaphragmed pith.

LEAVES Deciduous, alternate, simple, entire, obovate to elliptic, shiny, somewhat leathery, 3–10 cm long and 2–6 cm broad, turning pleasing shades of red and yellow in autumn.

FLOWERS Inconspicuous, greenish-white; petals very small and fleshy, or wanting.

FRUIT An oblong or ovoid, blue-black drupe, 8–12 mm long with a 10–12-ribbed stone (pit), and thin, acrid flesh, borne on pedicels 3–6 cm long.

WHERE FOUND Occasional in bogs but more common in low, acid woods, swamps, and on shores of streams and lakes.

NOTES The bright scarlet foliage is distinctive in autumn.

OTHER NAMES Black Gum, Pepperidge, Sour Gum, Tupelo, Tupelo Gum.

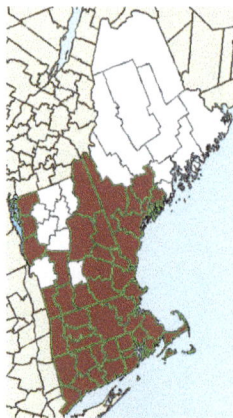

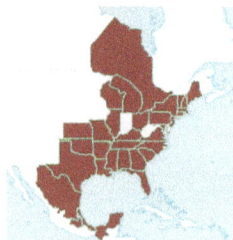

CURTIS HANSEN

BLACK TUPELO (*Nyssa sylvatica*)

JAKE OFLAHERTY

■ Black Ash

Fraxinus nigra Marsh.

GROWTH FORM Sparsely branched, small to medium-sized timber-tree, 15–25 m tall (smaller in bogs), with rather flaky or corky bark.

LEAVES Deciduous, opposite, odd-pinnately compound with 7–13 leaflets (the lateral ones sessile); a pad of tawny hairs extending across the rachis at the juncture of the leaflets; bruised foliage with the odor of *Sambucus* (Elder).

FLOWERS Polygamous or dioecious, naked, appearing in long panicles before the leaves emerge, April–May.

FRUIT Single samara, with nearly flat seed body completely surrounded by the wing.

WHERE FOUND Wet borders of bogs; also swamps and shores as well.

NOTES Generally less important as a timber-tree than other northern ashes because of the poorly shaped bole and inferior quality of the wood. However, from early times the wood has been used by American Indians in making pack baskets and other articles in the northeastern United States and Canada.

OTHER NAMES Basket Ash, Brown Ash, *Frêne noir* (Quebec), Hoop Ash, Swamp Ash, Water Ash.

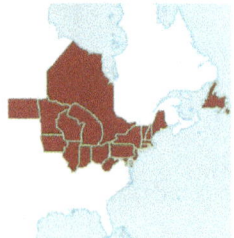

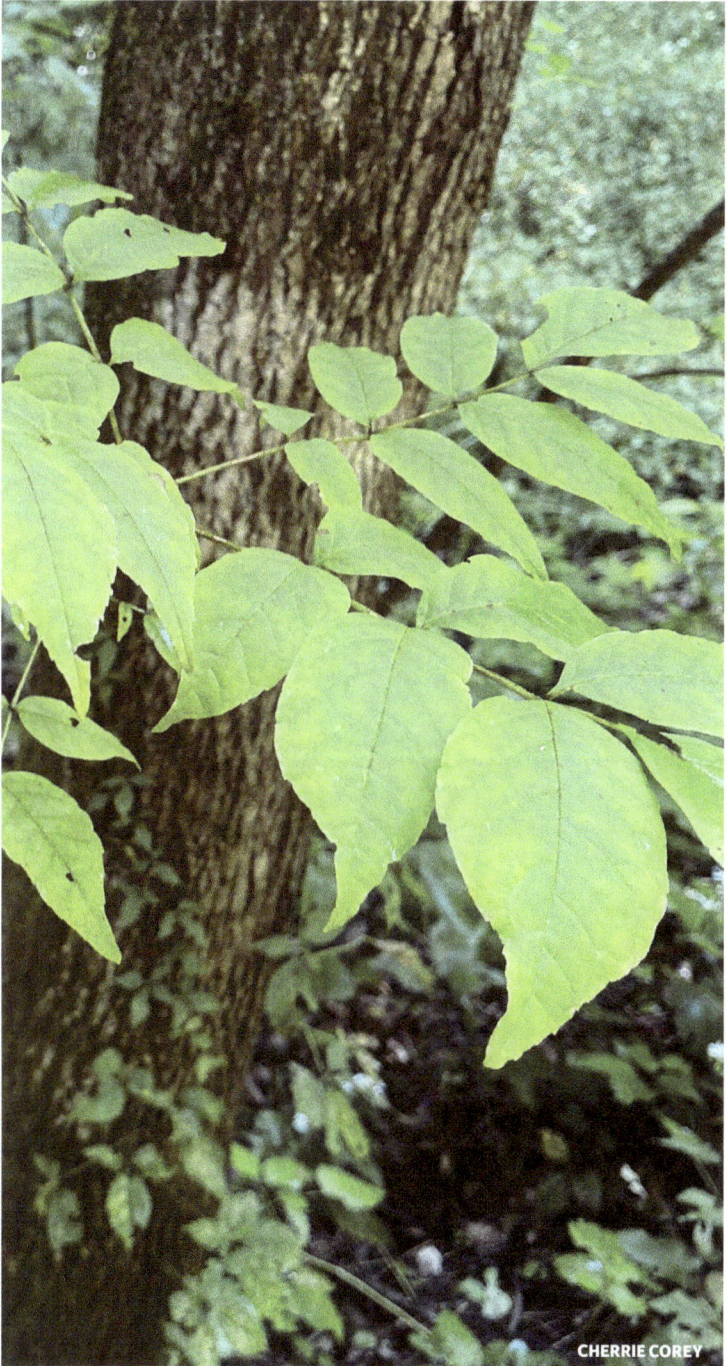

CHERRIE COREY

BLACK ASH (*Fraxinus nigra*)

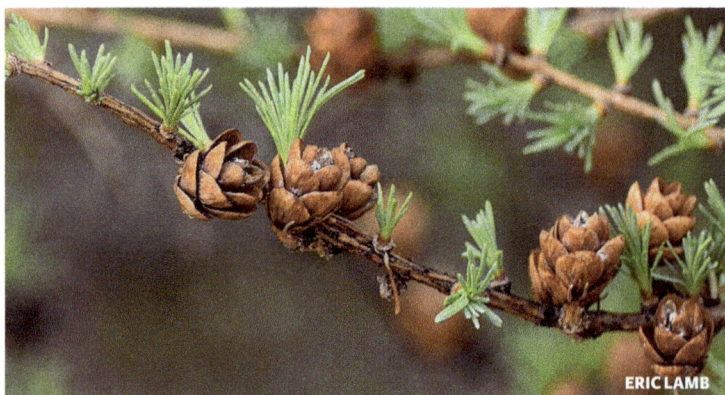

ERIC LAMB

■ Tamarack

Larix laricina (Du Roi) K. Koch

GROWTH FORM A deciduous conifer to 35 m high (much dwarfed, scraggly, and stunted in bogs), with open pyramidal crown and shallow, wide-spreading root system; bark scaly, gray to reddish brown; wood hard and resinous.

LEAVES Deciduous, simple, linear, soft, flexible, 1–2.5 cm long, arranged in clusters of many on the short spurs, spirally arranged on the long shoots of the season; 3-angled, turning yellow and falling from the branchlets in September to November.

FLOWERS The yellow staminate flowers globose to oblong, stalked or sessile, consisting of numerous short-stalked, spirally arranged anthers; the pistillate flowers subglobose, crimson or greenish with red bracts, April to May.

FRUIT Cones erect, 1–2 cm long, narrowly ovoid with visible bracts near the base, cone scales few, shiny, tan; cones persistent on the branchlets for several months or longer.

GROWTH FORM Found as a small, poorly developed tree in bogs chiefly on the hummocks where there is a thick accumulation of sphagnum; also along the wet margins where the sphagnum blanket is thinner and some mineral soil is available; common in swamps elsewhere where it becomes a fairly large tree.

NOTES Often (erroneously) called "juniper" (at least in northern Maine), and known by Indians in New York State as "Ka-neh-tens". Formerly used as "ship knees" in wooden vessels.

OTHER NAMES Alaska Larch, American Larch, Black Larch, Eastern Larch, *Épinette rouge* (Quebec), Hackmatack, "Juniper".

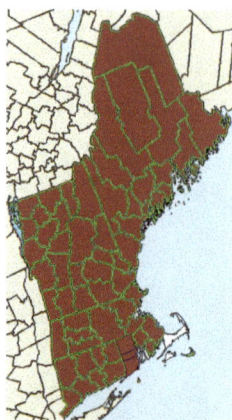

STEVEN LAMONDE

TAMARACK (*Larix laricina*)

JASON GRANT

■ Black Spruce

Picea mariana (P. Mill.) B.S.P.

GROWTH FORM Resinous evergreen conifer, up to 30 m high (much dwarfed in bogs where its lower branches often root in the sphagnum and perpetuate the tree in this manner); a ragged-appearing tree, forming a narrow, often irregular head; twigs pubescent.

LEAVES Evergreen, simple, bluish-green with whitish bloom, alternate, 6–13 mm long, needle-shaped, 4-sided, attached to the twigs by short, peg-like stalks which remain as rough projections when the leaves are ultimately shed (or specimens are dried).

FLOWERS Monoecious: male flowers catkin-like, dark red; female flowers cone-like, purplish, April–May.

FRUIT Ovoid, dull grayish or purplish-brown, blunt-pointed, scaly cones, 2–3 cm long, borne on short, curved stalks, maturing in 1 season but persisting for several years; cone-scales stiff, with erose or dentate margins, narrowed toward the tip.

WHERE FOUND Adjusting to bogs, as does Tamarack, where it occurs singly or in small isolated groups over wide areas; trees often distorted and stunted when growing in deep sphagnum but attaining better size and form near the wet bog borders; elsewhere on cool slopes and in humus-filled depressions at the higher elevations.

NOTES Resembles **Red Spruce** (*Picea rubens* Sarg.) and, at times, difficult to separate from it. Frequently hybridizing with red spruce, especially in the north.

OTHER NAMES Bog Spruce, Eastern Spruce, *Épinette noire* (Quebec), Shortleaf Black Spruce, Swamp Spruce.

BLACK SPRUCE (*Picea mariana*)

ALEXIS GODIN

■ Jack Pine

Pinus banksiana Lamb.

GROWTH FORM An evergreen conifer to 25 m high (on favorable sites) with spreading branches and resinous wood, but often shrubby and stunted in bogs where it is occasionally found.

LEAVES Evergreen, simple, needle-shaped, thick, short, divergent, sometimes twisted, 2–4 cm long, bright or dark green, borne 2 per fascicle bound together at the base by a persistent papery sheath.

FLOWERS Monoecious: male flowers (cones) yellow; female flowers (conelets) purple, May–June.

FRUIT A conical, oblong, or conic-ovoid, curved, serotinous, smooth, shiny, tawny or yellowish cone, 3–5 cm long, persisting on the branchlets (or branches) for several years; cone-scales thin, minutely spine-tipped.

WHERE FOUND In sparse sphagnum, where the tree is often stunted and crooked; elsewhere on barren, sandy, or rocky soil where it is of commercial importance, especially in the north where it covers vast areas not suitable for many other tree species.

NOTES Often planted in dry, sandy places where the soil is too poor for satisfactory growth of most other species.

OTHER NAMES Banksian Pine, *Cyprès* (Quebec), Gray Pine, Scrub Pine.

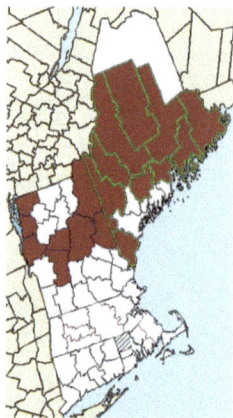

JYOUNG2399

JACK PINE (*Pinus banksiana*)

QUINTEN WIEGERSMA

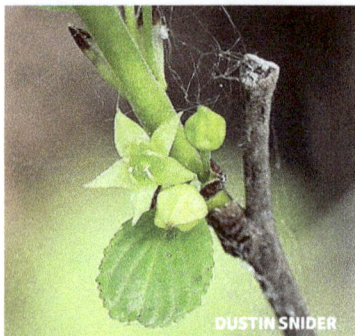
DUSTIN SNIDER

■ Alder-leaf Buckthorn

Rhamnus alnifolia L'Hér.

GROWTH FORM Low, upright or spreading shrub rarely exceeding 1 meter high, with few or no branches; young branchlets, minutely downy.

LEAVES Deciduous, alternate, simple, elliptic to oval or ovate, 4–10 cm long, unequally crenate-serrate with nearly straight veins.

FLOWERS Petals wanting, expanding with the leaves; sepals 5; May–July.

FRUIT Berry-like black drupe containing 2–4 seed-like, cartilaginous, 3-seeded nutlets, ripening in autumn.

WHERE FOUND Open places in the bog or where the sphagnous mat is thin; also swamps, low woods, and tall-grassy meadows.

NOTES Thrives well in calcareous areas.

OTHER NAMES None.

ALDER-LEAF BUCKTHORN (*Rhamnus alnifolia*)

ER-BIRDS

■ Black Chokeberry

Aronia melanocarpa (Michx.) Ell.

GROWTH FORM Similar to **Purple Chokeberry** (*Aronia prunifolia*) but usually lower in stature, with glabrous branchlets, 0.5–3 m high.

LEAVES Deciduous, alternate, simple, elliptic to broadly oblanceolate or obovate, acuminate at tip, lower surfaces (except the midrib) glabrous, crenate-serrate, 2–9 cm long and 0.5–4 cm broad, variable.

FLOWERS About 1 centimeter broad, with 5 white or pinkish-tinged petals, borne in several-flowered, glabrous inflorescences, April–early July.

FRUIT A small, subglobose or pear-shaped, glabrous, shining, black or black-purple, juicy, berry-like pome, 6–8 mm across, the tip creased in to form a 5-angled star, August–October.

WHERE FOUND Occasional in sphagnous areas in the bog; more common in peats, low, dry thickets and clearings, or on bluffs or ledges.

NOTES The flowers and (in the fall, the pinkish or old rose-colored leaves) add local color to the bog.

OTHER NAME *Gueules noires* (Quebec).

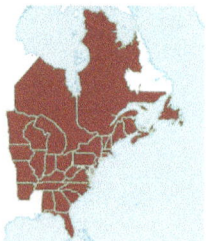

BLACK CHOKEBERRY (*Aronia melanocarpa*)

THOMAS KOFFEL

■ Purple Chokeberry

Aronia prunifolia (Marsh.) Rehd.

GROWTH FORM Slender colonial shrub 0.5–3 m high, with white-pubescent new branchlets, spreading by subterranean offsets.

LEAVES Deciduous, alternate, simple, crenate-serrate, broadly oblanceolate to narrowly obovate or sub-elliptic, abruptly acuminate at tip, dark green and glabrous with characteristic stalked, dark glands on the midrib above, grayish tomentose and pale beneath, 2–9 cm long and 0.5–4 cm broad, variable.

FLOWERS About 1 centimeter broad, with 5 white or pinkish-tinged petals, borne in several-flowered, pubescent inflorescences, April–early July.

FRUIT A small, subglobose or pear-shaped, dark purple or purplish black, juicy, berry-like pome, 8–10 mm in diameter, the tip creased in to form a 5-angled star, September–November.

WHERE FOUND Thriving in bogs, even in considerable depths of sphagnum; also low thickets and wet to dry soils.

NOTES The leaves turn pleasing shades of pink or old-rose, giving a local touch of color to the bog.

OTHER NAMES None.

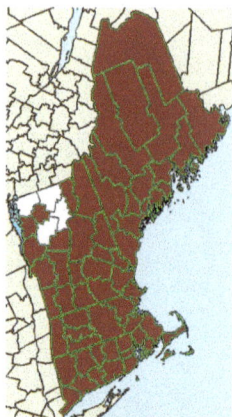

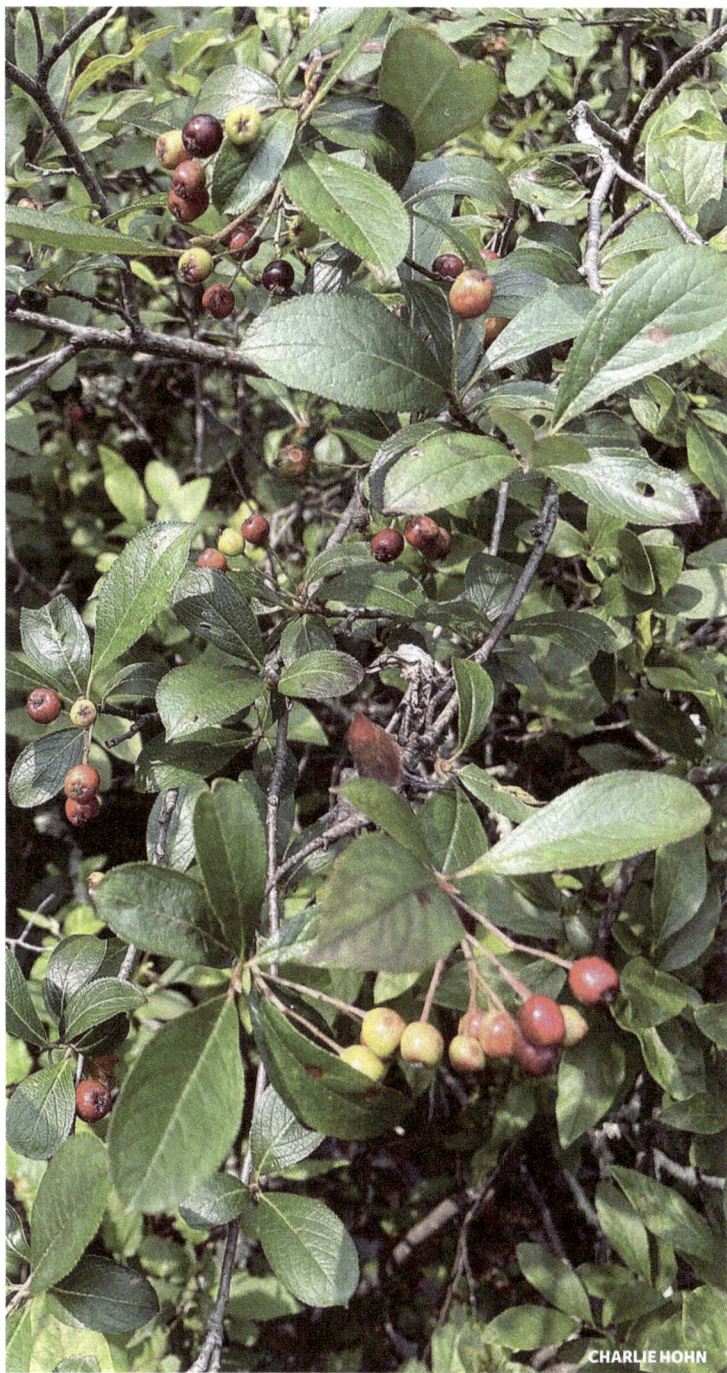

CHARLIE HOHN

PURPLE CHOKEBERRY (*Aronia prunifolia*)

SIGITAS JUZĖNAS

■ Shrubby Cinquefoil

Dasiphora fruticosa (L.) Rydb.

GROWTH FORM Shrub 0.2–1 meter tall with pale, outer shreddy bark.

LEAVES Deciduous, alternate, odd-pinnately compound with 5 or 7 entire, narrowly oblong to lanceolate or oblanceolate leaflets, with revolute margins, glabrous or densely white-villous, with persistent, sheathing stipules.

FLOWERS 1.5–3 cm broad, yellow, with 5 petals, June–October.

FRUIT Numerous small, dry achenes borne on a conical receptacle.

WHERE FOUND Occasional in bogs but more common on wet or dry open ground especially of calcareous origin.

NOTES Many horticultural forms known in the trade.

OTHER NAMES Golden-Hardhack, Widdy (Newfoundland). Formerly classified as *Potentilla fruticosa* L.

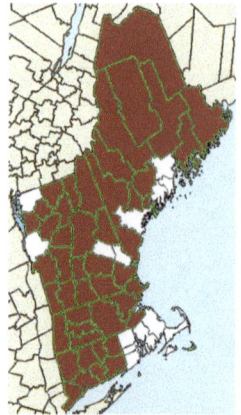

FRANCK CABOT

SHRUBBY CINQUEFOIL (*Dasiphora fruticosa*)

ERIK ERBES

■ Northeastern Rose

Rosa nitida Willd.

GROWTH FORM Low, slender, extremely bristly shrub 0.2–1 meter high, with canes (stems) 2–5 mm in diameter, from slender, stoloniferous rhizomes.

LEAVES Deciduous, alternate, odd-pinnately compound with 5–9 finely serrate, narrrowly elliptic or oblong-oval leaflets; stipules adnate to the petiole (a distinctive feature of roses).

FLOWERS Rather conspicous, pink, solitary or in bristly, few-flowered inflorescences; in evening with fragrance of Lily-of-the-Valley (*Convallaria*), June–August.

FRUIT Hairy achenes enclosed in a dark red, subglobose, fleshy receptacle or "hip".

WHERE FOUND Wet, open areas, often near the bog border; also wet thickets, pond margins, etc.

NOTES Occasional in bogs, lending local color and fragrance.

OTHER NAMES Shining Rose, Swamp Rose.

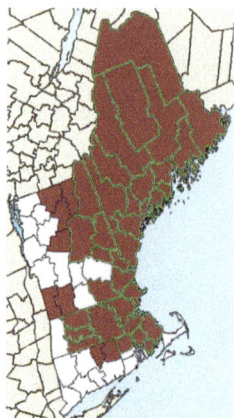

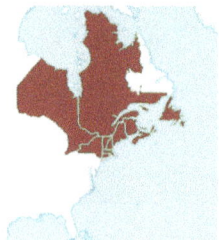

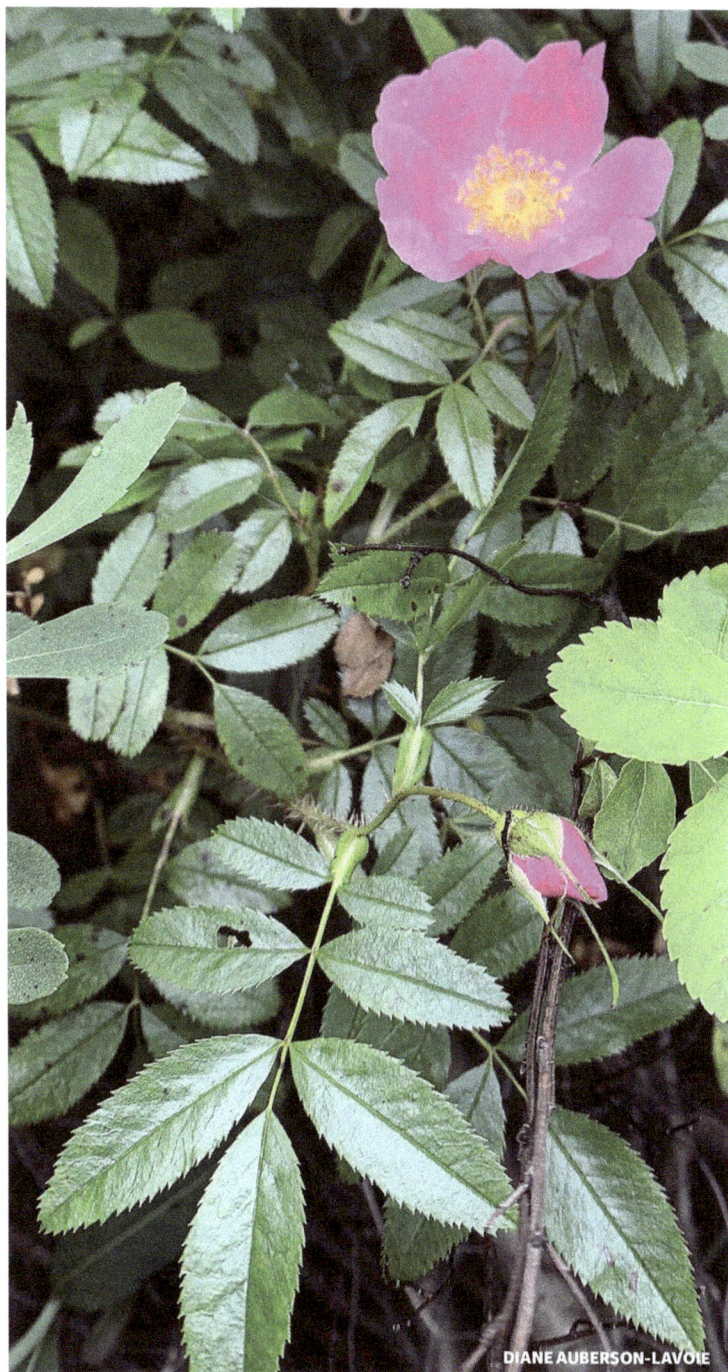

DIANE AUBERSON-LAVOIE

NORTHEASTERN ROSE (*Rosa nitida*)

OWEN STRICKLAND

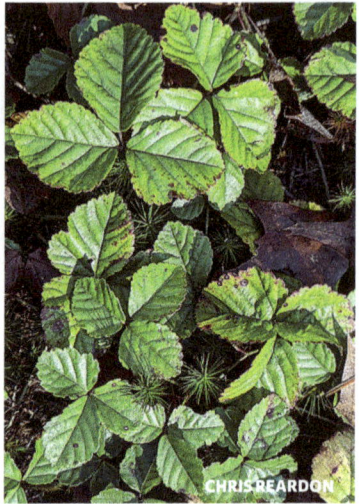

CHRIS REARDON

■ Swamp Dewberry

Rubus hispidus L.

GROWTH FORM Half-evergreen; canes (stems) to 2.5 m long and 5 mm thick, prostrate or trailing, with numerous mixed bristles and glands.

LEAVES Mostly evergreen (often partially browning during winter and spring), lustrous, dark green, often purple- or bronze-tinted below, alternate, compound with 3 (rarely 5) obtuse or short-pointed, obovate to rhombic-ovate or sub-orbicular, blunt-toothed leaflets; pedicels bristly.

FLOWERS White, with petals 5–12 mm long and less than half as broad, borne on pedicels 0.5 to rarely 4 cm long, flowering from late May to early September.

FRUIT An aggregate of purplish, tardily blackening, glabrous, sour druplets, 6–15 mm in diameter, ripening from mid-August to October.

WHERE FOUND Only occasional in bogs, usually in the less-wet areas, mostly near the margins; a common species of widespread distribution, more commonly found on wet or dry soil, in ditches, swales, and open woods.

NOTES Sometimes used as a ground cover; the semi-evergreen leaves and fruits are utilized by wildlife, especially the leaves, in winter when similar food ("greens") is not plentiful.

OTHER NAMES *Mûrier* (Quebec), **Swamp Black-berry**, **Bristly Dewberry**.

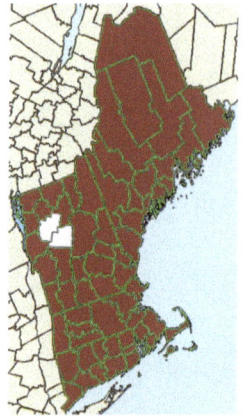

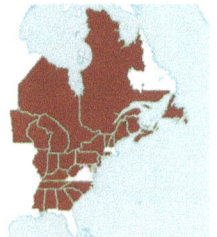

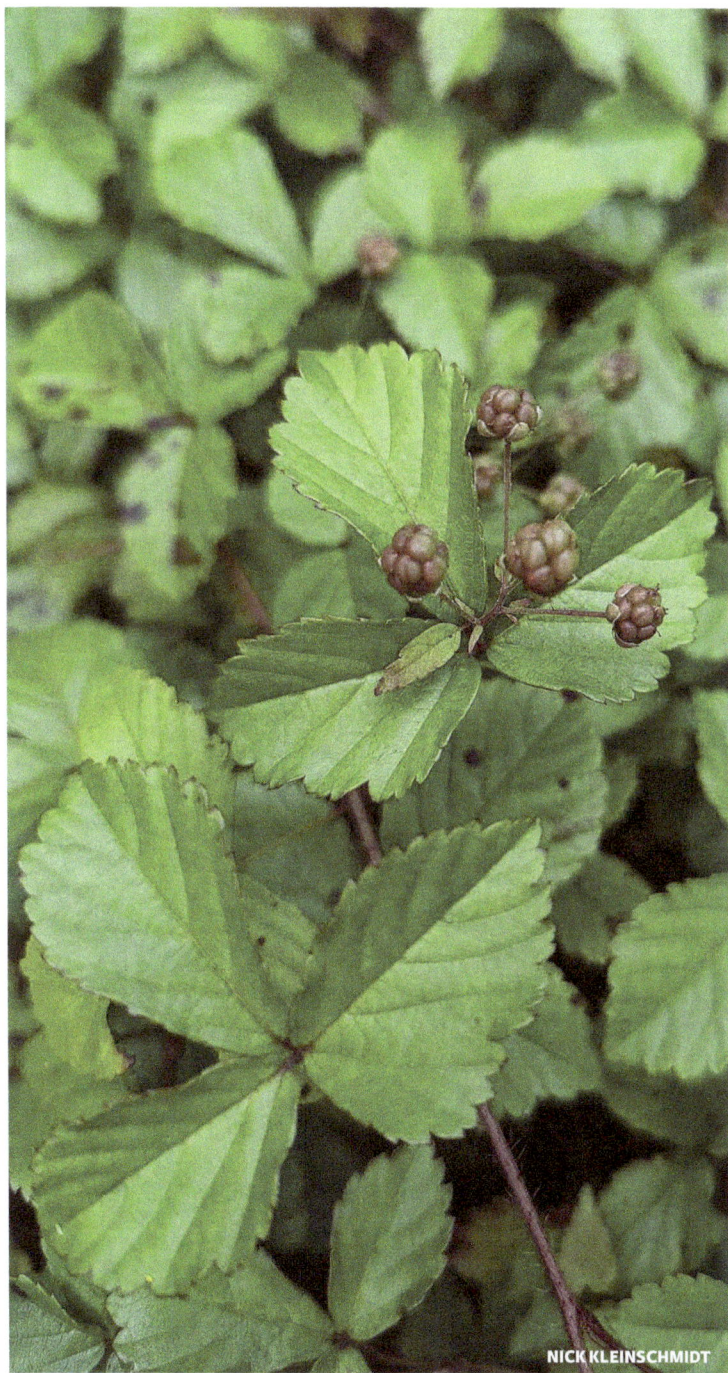

NICK KLEINSCHMIDT

SWAMP DEWBERRY (*Rubus hispidus*)

■ Meadowsweet

Spiraea latifolia (Ait.) Borkh.

GROWTH FORM Erect, wiry shrub to 1.5 m high with dark reddish-brown, glabrous, angled branches and branchlets.

LEAVES Deciduous, alternate, simple, obovate to oblanceolate, glabrous, thin, coarsely and often doubly serrate, slightly bluish beneath, mostly 1.5–7 cm long.

FLOWERS Inflorescences of broad, open, spreading, glabrous panicles 0.5–3 dm long; petals 5, white or slightly pinkish, June to September.

FRUIT Follicles 5–8, dehiscent along the inner suture with several tiny, oblong seeds, somewhat persistent in autumn.

WHERE FOUND Open places in the bog where the sphagnum is thin, or along the bog margins; also low ground or dry, open places elsewhere; not typically a bog plant but sometimes found there.

NOTES Attractive in flower and much sought by bees for nectar and pollen; sometimes a nuisance in hay fields and other places where the tough, wiry stems foul up the knives of the mowing machines.

OTHER NAMES *Thé du Canada* (Quebec).

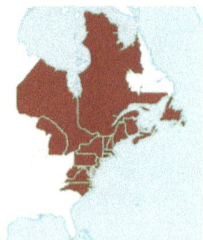

MARK EANES

MEADOWSWEET (*Spiraea latifolia*)

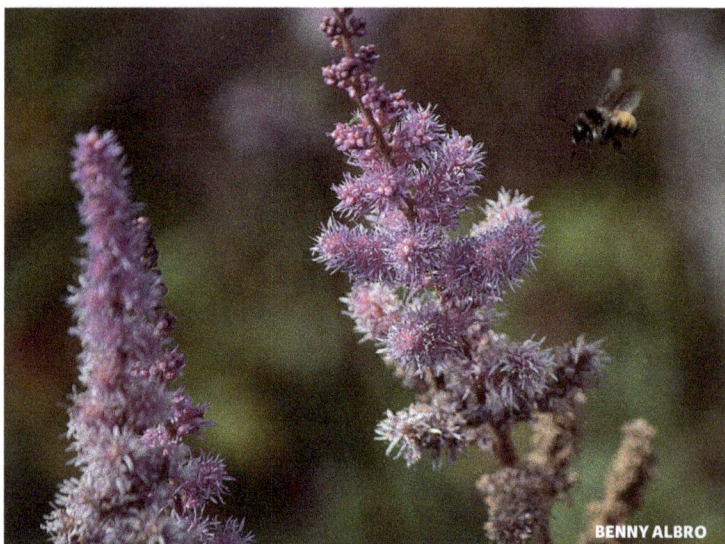

BENNY ALBRO

■ Hardhack

Spiraea tomentosa L.

GROWTH FORM Upright shrub to 1.5 m tall with angled, dark purplish branch-lets (but color often obscured by a coating of dense, matted brownish or tawny, woolly tomentum); branches tough and wiry.

LEAVES Deciduous, alternate, simple, unequally and often doubly serrate, ovate to ovate-oblong, rugulose above, densely woolly with yellowish or grayish tomentose hairs beneath, 3–7 cm long.

FLOWERS Inflorescences of closed, compact, narrow, spire-like brownish, tomentose panicles, 8–20 cm long; petals 5, deep rose or rose-purple (rarely white), showy, attractive, July–Sept.

FRUIT Follicles 5–8, dehiscent along the inner suture, with several, tiny, oblong seeds, somewhat persistent, autumn.

WHERE FOUND Open places in the bog and along the margins; more common in sterile, low grounds and pastures; not typically a bog plant but sometimes found there.

NOTES Attractive in flower and much sought by bees for nectar and pollen; sometimes a nuisance in hayfields and other places where the tough, wiry stems foul up the knives of mowing machines.

OTHER NAMES *Thé du Canada* (Quebec), **Steeple-bush**.

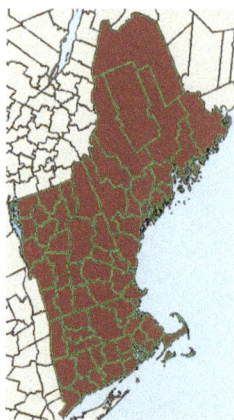

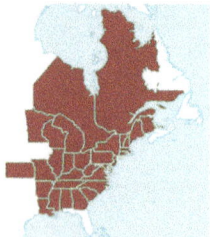

HARDHACK (*Spiraea tomentosa*)

JAY HORN

■ Buttonbush

Cephalanthus occidentalis L.

GROWTH FORM Smooth, low, wide-branching shrub to one or two m tall (in our area), often growing in water.

LEAVES Deciduous, opposite or whorled, simple, entire, ovate or elliptic-lanceolate, shiny, glabrous or somewhat pubescent, 6–15 cm long.

FLOWERS Small, creamy-white, sessile, borne in axillary, globose heads, on long peduncles, July–August.

FRUIT Heads of 2–4 closed, seeded nutlets, the heads falling apart and splitting from the base upward.

WHERE FOUND Open places in the bog, often permanently growing in water; also commonly in swamps, pond borders, and margins of streams.

NOTES A wide-ranging species becoming a small tree in the southwestern part of its range; conspicuous in bloom because of its "feathery" flowering heads.

OTHER NAMES *Bois noir* (Quebec), **Bush Globe-flower, Common Buttonbush.**

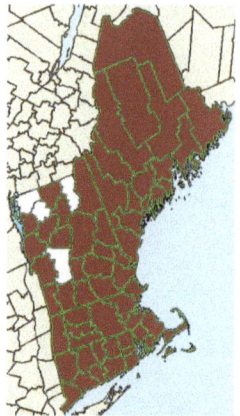

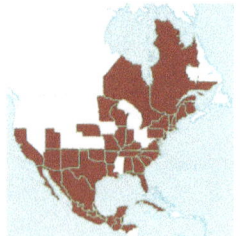

BUTTONBUSH (*Cephalanthus occidentalis*)

ROB FOSTER

NORMA MALINOWSKI

■ Bog Willow
Salix pedicellaris Pursh

GROWTH FORM Upright or slender, creeping and stoloniferous, looselybranching shrub to 1 meter tall, with glabrous, flexible, erect branches.

LEAVES Deciduous, alternate, simple, entire, linear-oblong to ellipticobovate, 1–5 cm long, revolute, without stipules, pale or whitened, glaucous and glabrous beneath.

FLOWERS In aments; the male and female sexes on different plants (dioecious), each with a gland at base, late April–July.

FRUIT A glabrous capsule, 5–10 mm long, June–July.

WHERE FOUND Open places in the bog; also sphagnous shores, occasionally subalpine.

NOTES One of the few species of local willows with entire (smooth-margined) leaves.

OTHER NAME *Saule* (Quebec).

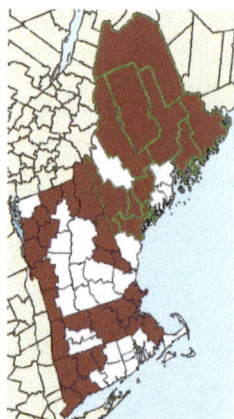

SARAH JOHNSON

BOG WILLOW (*Salix pedicillaris*)

NORMA MALINOWSKI

■ Dwarf-Mistletoe

Arceuthobium pusillum Peck

GROWTH FORM Tiny, fleshy to slightly woody plant, only 0.6–2 cm high, commonly parasitic on the branchlets of **spruce** (*Picea*) rarely on **tamarack** (*Larix*) and **pine** (*Pinus*); stems glabrous, olive to brownish or purplish, nearly terete or 4-angled in cross-section, simple or branched, brittle at base, arising from the cambium of the host and perennating (overwintering) in scattered clusters.

LEAVES Tiny, entire, leathery, appressed, scale-like, opposite, without stipules, connate, obtuse, suborbicular, olive or brown.

FLOWERS Dioecious, tiny, borne in the axils of the scales, April to June.

FRUIT Ovoid or ellipsoid flattened berry or drupe, 2–3.5 mm long.

WHERE FOUND Parasitic on the branches of *Picea, Larix* and *Pinus* (in Europe, on *Juniperus*); occasional and sometimes abundant, especially on **Black Spruce** (*Picea mariana*).

NOTES The only parasitic woody plant in our area; often causing "witches' brooms" on the coniferous host.

OTHER NAMES *Petit Gui* (Quebec).

NORMA MALINOWSKI

DWARF-MISTLETOE (*Arceuthobium pusillum*)

ANDREW CONBOY

■ Red Maple

Acer rubrum L.

GROWTH FORM Common forest tree with irregular or rounded crown, to 40 m tall (much stunted in bogs); branchlets and twigs red.

LEAVES Deciduous, opposite, simple, 3–5 palmately lobed, with sharp sinuses coarsely serrate nearly to the base, 6–10 cm long, turning brilliant shades of red to yellow in the fall (or even mid-summer), in sharp contrast with the green background of pine and spruce, enhancing the bog with unforgettable beauty.

FLOWERS Dark red to scarlet, appearing in early spring long before the leaves, sometimes injured by late spring frosts, March–May.

FRUIT Small, paired samaras ("keys") with pinkish or red wings, maturing early May to July.

WHERE FOUND Borders or open places in the bog; a common tree of varied habitats, including swamps and uplands.

NOTES Often planted as a street tree. The scarlet leaves add much color to the landscape in the fall.

OTHER NAMES *Plaine* (Quebec), *Plaine Rouge* (Quebec), **Scarlet Maple, Soft Maple, Swamp Maple, Water Maple, White Maple.**

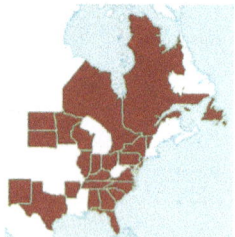

ASHWIN SRINIVASAN

RED MAPLE (*Acer rubrum*)

NICK KLEINSCHMIDT

■ Witherod

Viburnum cassinoides L.

GROWTH FORM Upright shrub 1–4 m tall with rather stiff branchlets and scurfy twigs.

LEAVES Deciduous, opposite, simple, oblong or lanceolate to oval or ovate, rather firm and leathery, with a crenulate, dentate, or partially entire margin, 2.5–15 cm long and 1.5–6 cm broad.

FLOWERS Ill-scented, white or yellowish-white, June–early August.

FRUIT Ellipsoidal to somewhat spherical drupe, changing from white through pink to blue-black, bloomy, 6–9 mm long; stone (pit) flat, el-liptic-oblong, September–October.

WHERE FOUND Borders and open places in the bog, where it survives as a much dwarfed plant; also thickets, clearings, swamps, and borders of woods.

NOTES Ripe pulp of fruit sweet; sometimes planted around the home and for landscaping.

OTHER NAMES *Alisier* (Quebec), **Wild Raisin**.

NICK KLEINSCHMIDT

WITHEROD (*Viburnum cassinoides*)

JUVIA HEUCHERT

■ Smooth Arrow-wood

Viburnum recognitum Fern.

GROWTH FORM Upright, bushy shrub 1–3 m high with gray bark, often forming clumps of several to many slender, straight, "arrow-like" stems; twigs angled, glabrous.

LEAVES Deciduous, opposite, simple, narrowly ovate to orbicular, glabrous, closely dentate with acute teeth, 3–9 cm long and 2–8 cm broad.

FLOWERS White, rather conspicuous, borne in glabrous, slender-stalked cymes, 5–8 cm across, mid-May to early July.

FRUIT A globose-ovoid, blue-black drupe, 6 mm long, late July–September.

WHERE FOUND Not strictly a sphagnous bog plant but occasionally invading the bog and persisting, usually as a depauperate plant; mostly along bog borders where some mineral soil is available; much more common and vigorous in moist or dry woods or thickets.

NOTES Adaptable and much used in the landscaping.

OTHER NAMES Laurestinus, Viburnum, *Viorne* (Quebec).

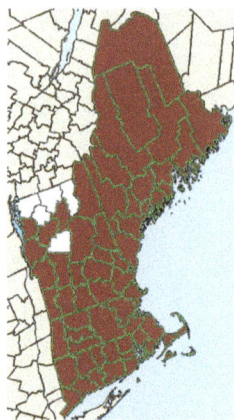

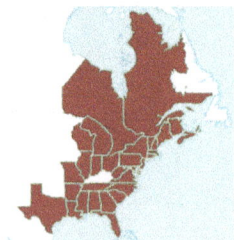

TANYA RISEMAN

SMOOTH ARROW-WOOD (*Viburnum recognitum*)

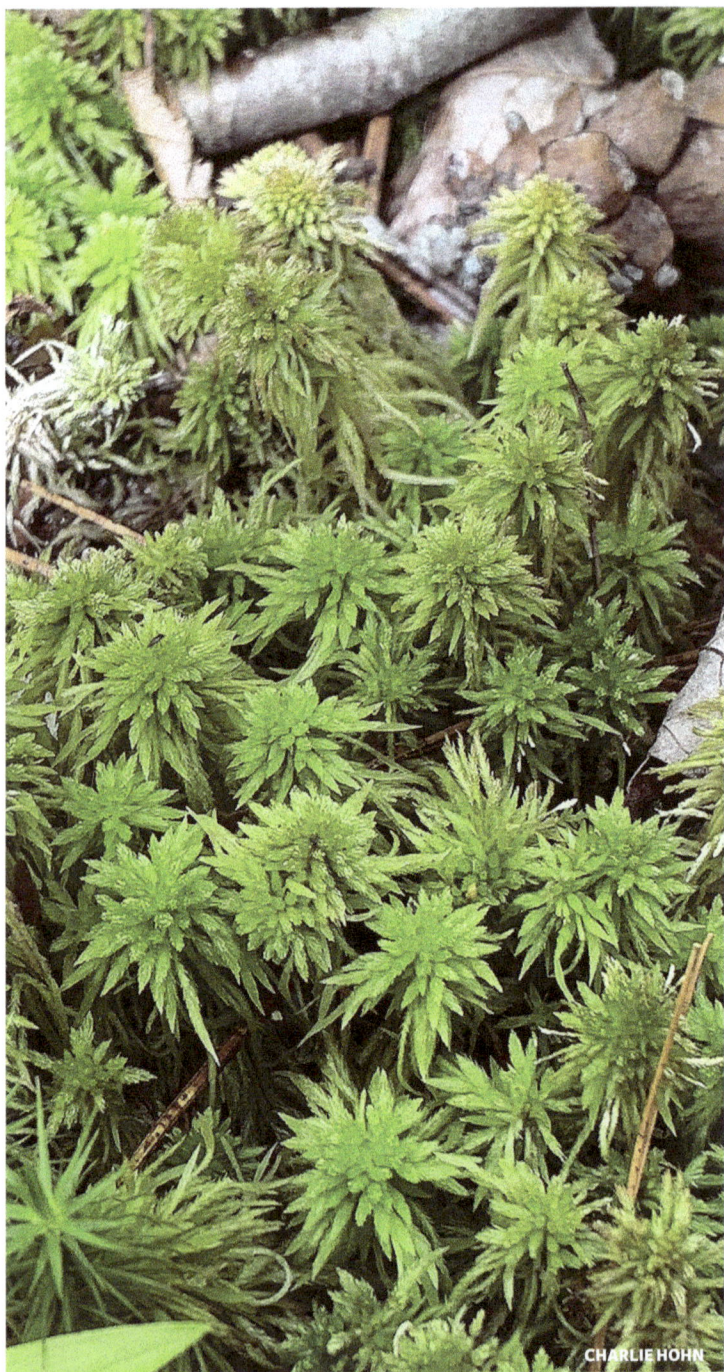

CHARLIE HOHN

SPHAGNUM MOSS (*Sphagnum* spp.)

Achene. A small, dry, hard, 1-celled, 1-seeded fruit which does not split open along definite lines at maturity.

Acuminate. Tapering to a long, sharp point.

Acute. Sharp-pointed, without tapering sides.

Adnate. United to an organ of a different kind.

Alternate. Said of leaves, buds, or leaf scars occurring singly at different points on the twig or branchlet.

Ament. Catkin, usually unisexual, often flexuous.

Angiosperm. Subdivision of seed-plants in which the ovule or ovules are enclosed in an ovary which becomes a fruit at maturity.

Anther. The pollen-bearing part of a stamen.

Axillary. Situated in the upper angle between bud or leaf scar and twig or branchlet.

Berry. A simple fleshy or pulpy fruit with usually many seeds. It contains no pit (stone) or core.

Bloom, Bloomy. With a whitish powdery covering, often waxy in nature.

Boss. A knob-like or rounded protuberance.

Bract. A somewhat modified leaf, occurring below and often belonging to a flower or inflorescence.

Bud. A rudimentary twig or growing point containing undeveloped vegetative or floral parts, often protected by scales.

Calcareous. Limy.

Callous. A hard protuberance or projection firmer than the surrounding tissue.

Campanulate. Bell-shaped.

Capsule. A dry, many-seeded fruit which splits open along several predetermined valves or cracks at maturity to liberate the seeds.

Ciliate. Fringed along the margin with hairs.

Compound. Said of a leaf which is divided into separate blades or leaflets.

Conifer, Coniferous. Cone-bearing woody plant.

Connate. United to another similar or like structure.

Corolla. A collective term for all the petals.

Crenate. Bearing rounded, often glandular, teeth.

Cuneate. Wedge-shaped, narrowly triangular; a term used to describe leaf bases.

Cyme. A broad or flat flower cluster with the central or terminal flowers blooming earliest.

Deciduous. Falling from a plant as the result of a natural process called "abscission"; not evergreen.

Dehisce, Dehiscent, Dehiscing. Opening along certain pre-determined lines or cracks in order to liberate the seeds.

Dentate. Toothed, with the teeth directed outward.

Diaphragmed. A solid pith having regularly-spaced horizontal partitions.

Dioecious. Unisexual, with the two kinds of flowers on separate plants.

Drupe, Drupaceous. A simple fleshy fruit with usually one or few stony pits. (The pit is not the seed proper but is the hard, stony inner fruit wall which surrounds the usually one seed.)

Elliptic, Elliptical. Tapering uniformly toward both rounded ends, broadest in the middle.
Entire. Without teeth or lobes of any sort.
Erose. With a jagged margin, as if weathered or eroded.

Ferruginous. Rust-colored.
Follicle. A simple, dry fruit opening only along one side to discharge its seeds.
Fruit. The seed-bearing structure of a plant.

Glabrate. Becoming glabrous (free from hairs) with age.
Glabrous. Smooth, not hairy.
Gland, Glandular. A secreting surface or structure.
Glaucous. Covered with a bluish-white or bluish-gray substance, often waxy in appearance.
Globose. Spherical.
Gymnosperm. A woody plant bearing naked seeds, without an ovary.

Heterogeneous. Not uniform in kind.
Homogeneous. All alike or of one kind.

Imbricate. Overlapping.
Inflorescence. The flowering part of a plant, especially the mode of its arrangement.

Lanceolate. Lance-shaped, several times longer than wide, broadest below the middle and narrowing to the apex.
Linear. Long and narrow, with nearly parallel margins.

Mesophyte, Mesophytic. Plants inhabiting medium or favorable sites and conditions as to moisture and light.
Midrib. The central or main rib of a leaf.
Mucronate. Ending in a short and small, abrupt tip.

Node. The point on the twig which normally bears one or more leaves.
Nutlet. A small, hard, dry, one-seeded fruit; a diminutive nut.

Oblanceolate. Lance shaped with the broadest part toward the apex.
Oblong. Two or three times longer than broad and with nearly parallel sides.
Obovate. Having the outline of a hen's egg, with the broader end toward the tip.
Obsolescent. Becoming rudimentary or extinct.
Obtuse. Blunt or rounded at the end.

Opposite. Two at a node directly across the axis from each other.

Ovary. The part of the flower (base of the pistil) which contains the ovules (potential seeds).

Ovate. Having the outline of a hen's egg, with the broader end toward the base.

Panicle. A loose or branched inflorescence with stalked flowers

Pedicel. The stalk or stem which supports a single flower.

Peltate. Shield-shaped and attached by the lower surface.

Petiole. Leaf stalk.

Pilose. Covered with soft hairs.

Pistil, Pistillate. The seed-bearing organ of a flower.

Pollen, Pollen grains. Spores or grains borne by the anther (swollen portion of the stamen), which produces the male gametes of the flower.

Polygamous. With perfect and unisexual flowers on the same or on different individuals of the same species.

Pome. A fleshy fruit, such as the apple; a core-fruit.

Prickle. A weak, slender, sharp outgrowth of the epidermis or bark.

Puberlous. Finely pubescent.

Puberulous, Puberulent. Minutely pubescent with short hairs.

Pubescent. Covered with fine, short, soft hairs.

Pulverulent. Appearing powdered with minute particles of dust.

Pyriform. Pear-shaped.

Rachis. The axis of an inflorescence or of a compound leaf.

Receptacle. The expanded or enlarged tip portion of the axis which bears the organs of a flower or collection of flowers.

Revolute. With margins rolled toward the lower side.

Rhizome. A prostrate or subterranean stem, usually rooting at the nodes.

Roseate. Rose-colored or rose-tinted.

Rugose. Wrinkled.

Rugulose. Shallowly or obscurely wrinkled.

Seed. The ripened ovule, consisting of embryo and seed coats, with or without endosperm (accumulated food).

Serotinous. Produced late in the season; as in Jack Pine, the cones remaining closed until opened by fire.

Serrate. Having sharp, forward-pointing teeth.

Sessile. Without a stalk.

Setose. Beset with bristles.

Shrub. A woody perennial, smaller than a tree, usually with several stems.

Simple. Said of a leaf with only one blade; not compounded into separate leaflets.

Sordid. Impure white, appearing dirty.

Stamen, Staminate. The pollen-bearing organ of a flower.

Stipule. An appendage at the base of the petiole or leaf or on each side of its insertion.

Stolon, Stoloniferous. A runner or any basal branch that is inclined to root.

Strigose. Beset with appressed, sharp, straight, and stiff hairs.

Strobile, Strobilus. An inflorescence composed of overlapping bracts or scales.

Style. The usually attenuated portion of the pistil connecting the stigma with the ovary.

Subglobose. Somewhat or slightly globose or spherical.

Suborbicular. Somewhat orbicular or circular.

Subshrub. A barely or only slightly woody plant or shrub.

Suture. A line or place of splitting.

Terete. Circular in cross-section.

Ternate. In threes.

Tetraploid. Having four times the basic (or twice the diploid) number of chromosomes.

Tomentose. Covered with dense, matted, woolly hairs.

Tomentulose. Minutely woolly.

Tomentum. A dense covering of matted hairs.

Tree. A woody plant potentially exceeding twenty feet in height and usually with a single trunk and a definite crown.

Urceolate. Urn-shaped.

Vein. Thread of fibrovascular tissue in a leaf.

Viscid. Glutinous, sticky.

Whorled. Occurring three or more at a node.

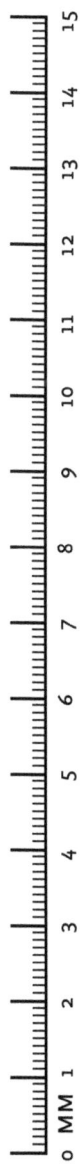

www.ingramcontent.com/pod-product-compliance
Lightning Source LLC
Chambersburg PA
CBHW052117030426
42335CB00025B/3023